Geography Pathways

Key Stage 3 for Northern Ireland

Year 8 Pupil's Book

Anne Fegan
Peter Corr
Stephen Roulston

Hodder Murray

A MEMBER OF THE HODDER HEADLINE GROUP

The Publishers would like to thank the following for permission to reproduce copyright material:

Photo credits p.7 Corbis/Alain Le Garsmeur; **p.10–11** *All* Stephen Roulston; **p.12–13** *All* Stephen Roulston; **p.15** from *Hiberniae Delineatio Atlas of Ireland* by Sir William Petty 1685 (this edition privately published by Frank Graham, 1968); **p.16 & 17** *t & b* Stephen Roulston; **p.18** Empics/Paul Faith/PA; **p.19** Corbis/Bob Krist; **p.20** Rex Features/Geoff Wilkinson; **p.21** *t* Stephen Roulston, *b* Tommy Fegan; **p.27** *l* Stephen Roulston, *r* Tommy Fegan; **p.28** Tommy Fegan; **p.29** *tl* Tommy Fegan, *cl* Courtesy of the Sprucefield Centre, *tr* Kevin Fegan, *bl & br* Stephen Roulston; **p.30** *t & b* Environment and Heritage Service; **p.31** *tl & tr* Environment and Heritage Service, *bl* Friends of the Earth; **p.33** Corbis/Mark Bolton; **p.35** Pacemaker Press International; **p.36** *All* Tommy Fegan; **p.37** *tl & tr* Corbis/Colin McPherson, *cr, cl, br & bl* Tommy Fegan; **p.49** Corbis/Alain Le Garsmeur; **p.50** Stephen Roulston; **p.52** *tl, tr & br* Stephen Roulston, *bl* Corbis/DK Limited; **p.53** *All* Stephen Roulston; **p.54** *t & bc* Stephen Roulston, *tc* PurestockX, *b* Corbis/Jim Sugar; **p.55** *tr, cr & br* Stephen Roulston, *cl* Getty Images/John Lawrence; **p.56** *l* Corbis/Henry Cabala/Beateworks, *r* Stephen Roulston; **p.57** Stephen Roulston; **p.58** Stephen Roulston; **p.59** *t* Stephen Roulston, *c* Caitlin Roulston, *b* Photo Library/Breck P Kent; **p.61** © Belfast Telegraph; **p.62** Empics/Rob Carr/AP; **p.68** *tl* Pacemaker Press International, *tr & bl* Empics/Paul Faith/PA, *br* Empics/Gareth Fuller/PA; **p.69** *tl* Empics/Paul Faith/PA, *tr & bl* Pacemaker Press International, *br* Empics/John Giles/PA; **p.74** *l* Pacemaker Press International, *r* courtesy of BBC Northern Ireland; **p.78** *t* Science Photo Library/Leonard Lessin, *cl* Science Photo Library/John Howard, *cr* Science Photo Library/Jerry Mason, *bl* Science Photo Library/Cape Grim B.A.P.S./Simon Fraser, *br* Science Photo Library/David Parker; **p.79–80** *Clouds* Photodisc; **p.80** Alamy/Jeff Morgan; **p.81** Greenmount Organic Farm; **p.82** *tl* Alamy/Barry Lewis, *tr* Alamy/Joe Fox, *cl* Corbis/Geray Sweeney, *cr* Corbis/Michael St. Maur Sheil, *bl* Alamy/Andrew Holt, *br* Geray Sweeney; **p.89** Greenmount Organic Farm; **p.90** Rex Features/Sipa Press; **p.91** Rex Features; **p.93** *l* M. Kirkwood/Drummuir Farm Ice Cream, *r* Alamy/South West Images Scotland; **p.94** www.beechdean.co.uk; **p.96** *All* Maurice Todd; **p.97** Getty Images/Xavier Bonghi; **p.98** Corbis/Roy McMahon; **p.100** Corbis/Mark E.Gibson; **p.101** *t* Corbis/Macduff Everton, *b* Corbis/Lester Lefkowitz; **p.102** *All* PurestockX; **p.104** *tl* Corbis/Image100, *tr* Popperfoto, *bl* Getty Images/Lambert/Hulton Archive, *br* Getty Images/Hulton Archive; **p.106** *tl* PurestockX, *tr* Alamy/Simon Rawles, *bl* Empics/John Kuntz/AP, *br* Ecoscene/Stuart Baines; **p.107** *tl* The Fair Trade Foundation, *tr* Alamy/Marl Layton/Travel Ink, *bl & br* PurestockX; **p.108** Stephen Roulston; **p.111** *l* PurestockX, *r* The Fair Trade Foundation; **p.120** Empics/Paul Faith/PA; **p.122** Tommy Fegan; **p.123** *t* Corbis/Galen Rowell, *l* PurestockX, *r* Getty/Sajjad Hussain/AFP; **p.126** *t* Robert Harding/Charles Bowman, *b* Geray Sweeney; **p.127** Corbis/Richard Cummins; **p.128** *t* Getty Images/Sean Sexton, *bl* Getty Images/William Lawrence/Sean Sexton, *br* Getty Images/Terry Vine; **p.129** Rex Features; **p.130** *All* Stephen Roulston; **p.131** Stephen Roulston.

Acknowledgements p.39 DETI statistical data in the table, Crown copyright material is reproduced with the permission of the Controller of HMSO; **p.70** British Broadcasting Corporation (BBC) (BBC Information), www.yahoo.co.uk copyright © 2007 Yahoo!, data supplied by the Met Office © Crown copyright, 2007; **p.72–3** data supplied by the Met Office © Crown copyright, 2007; **p.85 & 87** DARD statistical data from their website in four tables and graph, Crown copyright material is reproduced with the permission of the Controller of HMSO; **p.95, 136–7** maps reproduced from the 1993 and 2003 Ordnance Surveys of Northern Ireland Coleraine and Ballycastle maps with the permission of the controller of Her Majesty's Stationery Office, © Crown Copyright 2007, permit number 70023; **p.112** © 2007 Oxfam GB; **p.113** *The Guardian* copyright Guardian News & Media Ltd 2004; **p.120** *The Irish News*; **p.121** *The Belfast Telegraph* for the article 'Marine food chain link being affected by global warming', 21 November 2005, © *The Independent*.

Every effort has been made to trace all copyright holders, but if any have been inadvertently overlooked the Publishers will be pleased to make the necessary arrangements at the first opportunity.

Although every effort has been made to ensure that website addresses are correct at time of going to press, Hodder Murray cannot be held responsible for the content of any website mentioned in this book. It is sometimes possible to find a relocated web page by typing in the address of the home page for a website in the URL window of your browser.

Hodder Headline's policy is to use papers that are natural, renewable and recyclable products and made from wood grown in sustainable forests. The logging and manufacturing processes are expected to conform to the environmental regulations of the country of origin.

Orders: please contact Bookpoint Ltd, 130 Milton Park, Abingdon, Oxon OX14 4SB. Telephone: (44) 01235 827720. Fax: (44) 01235 400454. Lines are open 9.00–5.00, Monday to Saturday, with a 24-hour message answering service. Visit our website at www.hoddereducation.co.uk

© Peter Corr, Anne Fegan and Stephen Roulston 2007
First published in 2007 by
Hodder Murray, an imprint of Hodder Education,
a member of the Hodder Headline Group, an Hachette Livre UK Company.
338 Euston Road
London NW1 3BH

Impression number   5 4 3 2
Year   2011 2010 2009 2008 2007

Cover photos: *Giant's Causeway, County Antrim* © Hubert Stadler/CORBIS; *Milking Holstein Cattle* © Michael St. Maur Sheil/CORBIS; *Ceramicist working on vases at Belleek Pottery Ltd.* © Tim Thompson/CORBIS; *Fishing on Lough Erne* © Geray Sweeney/CORBIS; *Stone circles in the snow* © Geray Sweeney/CORBIS.
Illustrations by Mike Parsons, Barking Dog Art and Richard Duszczak, Cartoon Studio Limited
Designed in Palatino 11/13pt by Lorraine Inglis Design
Printed in Italy

A catalogue record for this title is available from the British Library

ISBN-13: 978 0340 900 109

CONTENTS

Welcome to the world of Geography!

Welcome to your new Geography course! This course is unique in that it has been written especially for pupils like you, living and going to school in Northern Ireland. You will find that you are investigating lots of **issues** you never thought of as 'Geography'. You will be actively asking questions, thinking through ideas, solving problems and making decisions about things that concern you and others in the world around you.

This book is for Year 8. It is the first in a three-book course for Key Stage 3. The main objective for this book is to build your awareness of your **local** area by investigating geographical issues that relate to Northern Ireland. There are seven chapters in this book. Each chapter introduces you to skills and concepts by exploring a particular geographical theme or topic.

Chapter	Description
I What is Geography?	Of course you know it's not *just* all mountains and rivers!
2 Ecosystems and people	How does your garden grow? This chapter on ecosystems also looks at how we affect the natural environment and why we should care about it.
3 The changing world of work	Why is work called 'economic activity'? This chapter also looks at change, including the challenges of living in a multi-ethnic and multicultural world.
4 Northern Ireland rocks!	What's under the ground and how does this affect the landscape?
5 Rain, rain, go to Spain	It won't! This chapter is all about our 'lovely' weather and how it affects our lives.
6 Farming today	This chapter looks at how farming has changed in Northern Ireland.
7 How much do we know about our food?	How far has some food travelled before it gets to you and what happens to it?

In addition, you will find other features at the end of your book, such as GIS and enquiries to cover the coursework and fieldwork elements of your course. All of these will require the demonstration of skills and the application of knowledge and understanding about Geography. Your teacher will guide you on how you will study them. In each chapter, you will find:

- **Learning intentions** which tell you the skills and knowledge you will be learning in the chapter.
- **Get Active** exercises which help improve your thinking and practise your geographical skills.
- **The Big Task** which pulls together all your work at the end of the chapter and gives you the opportunity to reflect on your own performance.
- **Key words** which are in bold and are explained in a glossary at the back of the book.
- **Thinking skills and personal capabilities icons** which show you at a glance where you have the opportunity to develop particular cross-curricular skills. These icons are explained in the table on page 5.

Thinking skills and personal capabilities

Skill	Icon	Description
Managing information		Research and manage information to investigate geographical issues; including identifying, collecting and using primary data and accessing and interpreting a range of secondary sources such as maps, text, images, charts and statistics, newspapers and internet sources.
Thinking, problem solving, decision making		Show deeper geographical understanding, be more critical, think flexibly, solve problems and make informed decisions.
Being creative		Demonstrate creativity and initiative when developing own ideas; communicate effectively in oral, visual, numerical and ICT formats showing clear awareness of audience and purpose.
Working with others		Work effectively with others in a group or team.
Self-management		Demonstrate self-management by working independently to manage, evaluate and improve own learning and performance.

Throughout the course, you will also discover how your investigation of geographical issues helps to develop your understanding of the 12 key elements listed in the table below. These are covered in all the subjects in the curriculum at school. The table shows a brief description of each key element. Note that not all the key elements are covered in this book. Some will be covered in Years 9 and 10.

Key element	Description
1 Personal Understanding	• Develop a sense of place and belonging. • Demonstrate an awareness of relationships to other places, peoples and environments.
2 Mutual Understanding	• Challenge stereotypes and perceptions of different places and people. • Investigate the impact of diversity on local settlement. • Investigate the factors that result in people having to make life-changing decisions.
3 Personal Health	• Investigate factors that affect personal health locally. • Explore the influences on global patterns of health.
4 Moral Character	• Challenge distorted views with appropriately balanced responses. • Take responsibility for choices and actions.
5 Spiritual Awareness	• Respond to the diversity and beauty of the natural and human world. • Reflect on experiences of a visit to a dramatic landscape.
6 Citizenship	• Investigate differences in lifestyle within and between countries. • Explore how we can play a role in helping to promote a fairer world.
7 Cultural Understanding	• Develop an understanding of how people interact with their natural environments.
8 Media Awareness	• Investigate the causes and consequences of an environmental event making the news and evaluate how it is reported in the media.
9 Ethical Awareness	• Research and debate ethical issues in Geography.
10 Employability	• Investigate how the skills developed through Geography will be useful in a career.
11 Economic Awareness	• Investigate the impact of globalisation and how it has produced winners and losers.
12 Education for Sustainable Development	• Investigate how physical processes operate to create distinct and diverse environments and the impact of conflict between social, economic and environmental needs. • Explore how we can exercise environmental stewardship.

The World Wide Web is a wonderful and rich **source** of information and you will find it very useful when you are using this book. There are a number of things that you need to be aware of when you are using the web.

1 Are you safe?

Never give out your personal details on the web, especially your full name, address, phone number or even your email address. If you see or read anything on the web that you feel uncomfortable about, show it to a teacher or to a parent or other trusted adult. The websites below give a lot of good information about this:

- www.urzone.com/advicezone/advice_internet.htm
- www.thinkuknow.co.uk.

Sometimes using search engines designed for younger users of the web will get better results: www.askforkids.com, www.yahooligans.com or www.kidsclick.org are all useful.

2 Is the stuff you find true and accurate?

There are no rules about what people put up on the web, so some of what you find there is wrong. There are a number of things you can do to be sure that the facts you have found are correct. Find out the type of person who has written it. Is the website an educational site? It might have .edu at the end (if it is from the USA) or .ac (if it is a UK site). You have to be careful about sites that have web addresses like www.stanford.edu/~jsmith. These *seem* to be educational sites, but the ~ means that this is the private page of a member of staff of that university or school and whatever is on it might be just that person's opinion.

What you find may be correct but may be out of date. Check if a date is provided on the site.

Double-check the information. You could check if the same information is on a lot of other sites. However, even if it is on a lot of sites, that still does not make it true!

3 Can you use what you find?

Just because it is on the web does *not* mean that you can use it, such as in a digital presentation or in a word processed document. Most of the content on the web is **copyrighted** (photographs, diagrams, cartoons, text, graphics, video or anything digital). This protects the person who created the content from people who could steal it and use it without their permission.

Many sites allow content to be used in schools. Check if that is the case. If there is no statement giving permission on the site, you could try emailing the site owner and asking for permission – most of the time they will be happy for you to use it. But if you are not given permission you shouldn't use it!

1 What is Geography?

This photograph was taken in County Down and shows typical northern Irish countryside

Learning intentions

In this chapter, I am learning:

- what Physical and Human Geography is about
- the origins of place names in Northern Ireland
- how to describe landscapes
- how to read photographs and use them to describe places.

Geography is all around you!

Geography is the story of *us* in the natural world that we inhabit. Geography is also the study of the Earth and the relationships that exist between the physical world and all of humanity.

- **Physical Geography** is the study of the natural world: the rocks, soil, **climate**, rivers, plants and animals. It is also about all the spectacular hazards and disasters that you often hear about in the news: **earthquakes**, volcanoes, storms, tsunamis, floods.
- **Human Geography** studies the behaviour of human beings, and how they interact with the natural world. It is about where and how people use the natural world for food, to live, work, and have fun. Human Geography often focuses on issues and problems that concern how we live, what mistakes we make and how we can correct them.

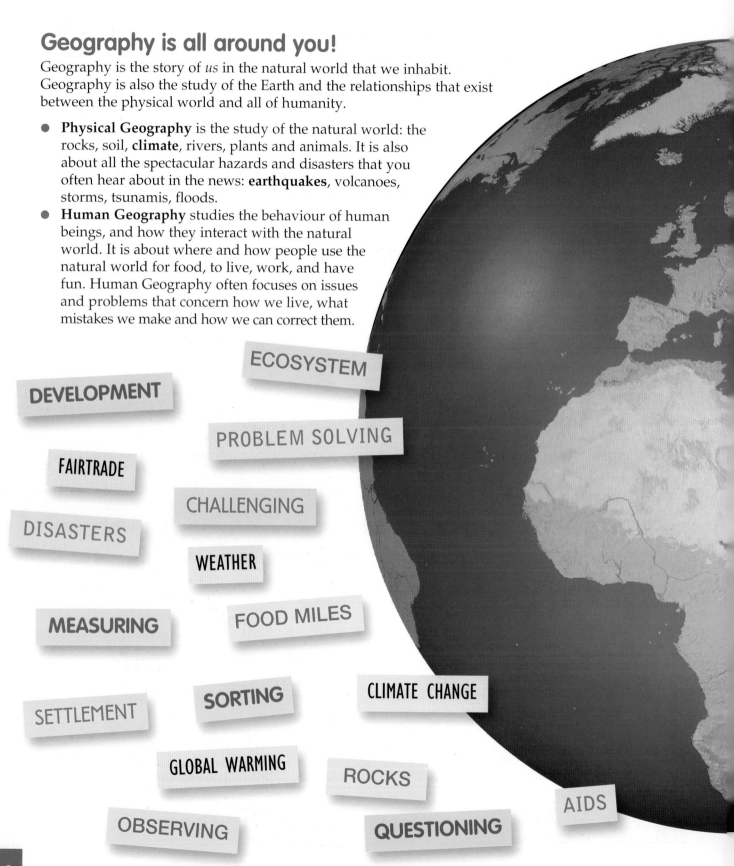

ECOSYSTEM

DEVELOPMENT

PROBLEM SOLVING

FAIRTRADE

CHALLENGING

DISASTERS

WEATHER

MEASURING

FOOD MILES

SORTING

CLIMATE CHANGE

SETTLEMENT

GLOBAL WARMING

ROCKS

AIDS

OBSERVING

QUESTIONING

Geography is a fascinating subject because it describes our Earth and asks us to question the world around us. Who are we? Why do we live where we do? How do we affect the **environment**? Are the decisions that we make about our lives successful?

This is a special book because it is about *your* Geography! It focuses on your place and your local area: Northern Ireland. It will allow you to get to know and to ask questions about what is going on here, what decisions need to be made and how to solve problems that affect you. It is part of your life story.

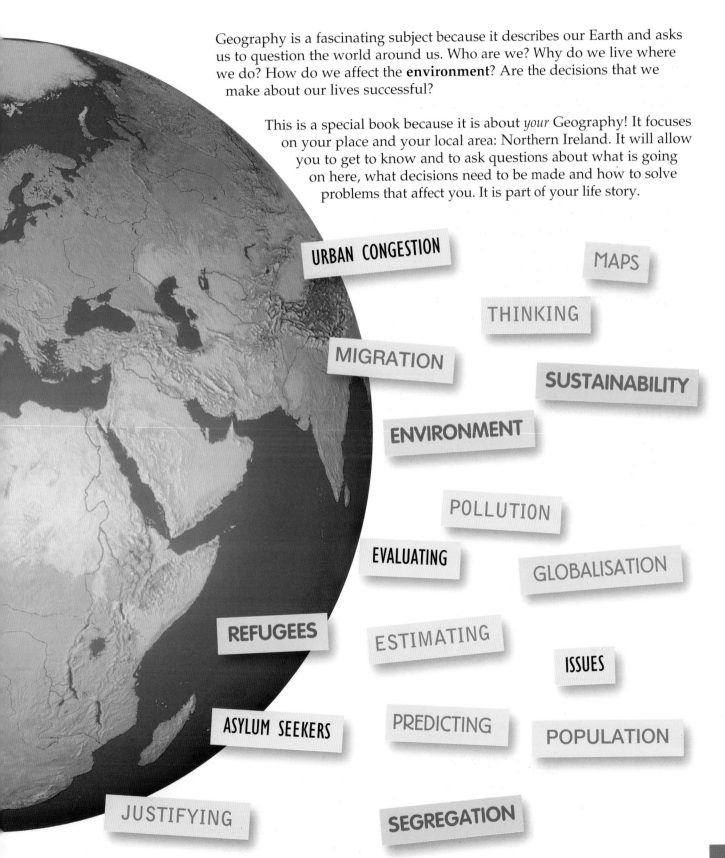

URBAN CONGESTION

MAPS

THINKING

MIGRATION

SUSTAINABILITY

ENVIRONMENT

POLLUTION

EVALUATING

GLOBALISATION

REFUGEES

ESTIMATING

ISSUES

ASYLUM SEEKERS

PREDICTING

POPULATION

JUSTIFYING

SEGREGATION

Where do you live?

Northern Ireland is a small area, part of the island of Ireland, and part of the United Kingdom. It is about 14,000 square kilometres in size. It has a lot of different types of scenery. There are also cities, towns and villages. There are motorways and small country lanes. It is difficult to show what a place like Northern Ireland is like by picking just a few photographs but here are some examples.

Armagh, County Armagh

Irvinestown, County Fermanagh

Enniskillen, County Fermanagh

Scenes of Northern Ireland

The Northern Ireland Tourist Board often picks the Giant's Causeway or White Park Bay for their website to show what Northern Ireland is like. If you were working for the Northern Ireland Tourist Board, what images would you put on the website? You can only choose five.

a) Using the **internet** to research, print out 20 images of Northern Ireland. Working with a partner, put them into three piles: ones you would like to use, ones you would not use and ones that you might use.

b) Divide a large sheet of paper into two with a line down the middle. Stick the pictures you think show Northern Ireland best on one side and those that don't fit your idea of Northern Ireland on the other.

c) Compare them to each of the other groups:
  • Did you pick the same as the group next to you?
  • If there were differences, why was that?
  • Do different people have different views as to what Northern Ireland is like? Is that surprising?

Fair Head, near Ballycastle, County Antrim

Belfast, County Antrim

What's in a name?

Irish or English place names?

These signs all show names of places in Northern Ireland. Some names are so old that we don't know when they were first invented and sometimes we cannot work out what they mean.

Most of us speak English today, but when you mention a place name, it is likely that you are speaking Irish! Most of our place names are in Gaelic, or Irish, as that was the main language spoken here in the past. Some place names come from other languages such as Ulster Scots too. So although you might know how to pronounce a place name, you might not know what the place name actually means.

Studying place names can tell you a lot about the place where you live and what it used to be like when it was first given its name.

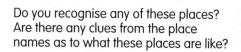

Do you recognise any of these places? Are there any clues from the place names as to what these places are like?

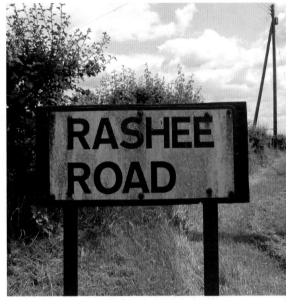

Townlands

One type of place name is only found in Ireland: the townland. Townlands are nothing to do with towns like we have today. Townlands were areas of **farmland** that had a cluster of farmhouses in the centre (the 'town'). If you live in the country you may know the name of your townland, but, even in towns and cities, there are still townland names used. Many townlands were given the Gaelic word Baile (pronounced Bally), meaning 'town', and this is a common part of Irish place names today.

Townlands such as Shankill (old church), Cloughfern (stone of the alder, a type of tree which grows near or in water), Garvaghy (rough field) and Ballygowan (townland of the smiths, people who work with metal) all tell us something about the history or the land after which the townland was named.

Gaelic spelling	How it usually appears now	Meaning	Example
Ard	Ard	Height	Ardglass (green height)
Baile	Bally	Town	Ballynahone Beg (little town of the river)
Bán	Ban, bane	White	Lurganbane (white ridge)
Beag	Beg	Small	Mullaghbeg (small hill)
Bearnas	Barnish, barnes	Gap in mountains	Barnish (gap)
Carraig	Carrick	Rock	Carrickmore (big rock)
Clár	Clare	Plain (flat land)	Clare (the plain)
Cnoc	Knock	Hill	Knockbane (white hill)
Dearg	Derg	Red	Aghaderg (red field)
Doire	Derry	Oak grove	Derrylough (the oak grove of the lake)
Dubh	Doo, duff	Black	Lough Doo (black lake)
Fada	Fad, fada	Long	Killyfad (long wood)
Garbh	Garvey, garvagh	Rough	Garvaghy (rough field)
Glas	Glass	Green	Moneyglass (green thicket)
Inis	Inish	Island	Inishbeg (small island)
Machaire	Maghera	Plain	Maghera (plain)
Mór	More	Big	Tamnaghmore (the big field)
Ramhar	Rammer	Broad	Drumrammer (broad ridge)
Ráth	Rath	Fort	Rathfriland (Fraiole's fort)
Saileach	Sally	Willow	Drumsallagh (ridge of willow)
Sliabh	Slieve	Mountain	Slievenabrock (mountain of the badgers)
Sruthán	Struan, surhan	Stream	Surhanleanantawey (stream of the sorrel meadow)
Tamlaght	Tamlacht	Pagan burial place	Tamhlacht (pagan burial place)
Tobar	Tober	Well	Tobernaveen (well of the Fianna)
Tulaigh	Tully	Mound or hill	Tullyree (hill of heather)
Tír	Tyr, tir	Land	Tyrone (Owen's land)

Some common parts of Irish place names

Get Active 1.2

a) Find the names of some place names (which might be townlands) close to you.
You could look on the internet to find what the names might mean.
A good starting point is: www.ulsterplacenames.org/celebrating_ulster's_townlands.htm.

b) Look at the place names in the 1:50,000 map on page 136. In squares 8331 and 8635 are two names of places and in 8336 there are two more. Which ones are townlands and which ones are more modern names?

c) Some parts of Irish place names are easy to remember. 'Mór' (usually changed to 'more' in the English translation) means 'big', and 'Beag' (usually changed to 'beg' in English) means 'small'. There is also 'Gort', which means 'cultivated field', 'mullagh' means 'hill', 'carn' means 'pile', and 'tober' means 'well'. Are there any place names on the map on page 136 which use these? Find them and give their grid references.

d) Are the names of the places in *your* area modern or old?

e) Look at the old map here that was published by Sir William Petty in 1685.
Find any place names that were the same, or very similar, compared to a modern map. Are there any other names you recognise on the old map? (Notice that they use 'ʃ' instead of 's' in the middle of names and that they abbreviate 'Bally' to 'B:'.)

f) Some old place names are being lost and replaced by modern names. Do you think this matters? Why?

A facsimile from Sir William Petty's *Atlas of 1685*

Reading photographs

Landscape is about the shape of the Earth, the plants that grow there, the soil, the mountains, the rivers, and the buildings and people too.

What can you see?

The landscape of the north-west of Northern Ireland is mostly flat. Farmland and houses extend to the foot of the mountain, known as Benevenagh. The sides of the mountain, where farming would not be so **productive**, are cloaked in **coniferous trees** which are harvested for **timber**. The top of the mountain is flat and has no buildings.

This photograph shows a **city** landscape but, even here, you can see fields and mountains in the distance. There are also rivers, streams and patches of **vegetation** in the centre of cities if you look for them. This is Belfast. The site of Belfast is nestled in a valley near the mouth of a river with many crossing points that attract travellers and so bring money into the city. What is your local town or city like?

Ballycastle is an example of a **coastal** landscape in County Antrim. There are some houses close to the sea. Many of these are owned by local people but some will be rented out as holiday homes to tourists.

The Sperrin Mountains in Northern Ireland is a **mountainous** landscape. This landscape has changed as people have abandoned small farms.

Get Active 1.3

What is the landscape like where you live? Find a postcard of it, or take a photograph, and put notes around it that describe the kind of landscape you can see, so that someone from another country would understand what your landscape is like. You will probably need to use geographical words, such as those in the glossary in this book, to describe your landscape properly.

Can you tell me the story?

Photographs arouse our interest and our curiosity and we ask questions and try to find out more. We *read* the photograph and try to find the story.

Every picture tells a story. Where did this happen? When? Who are the people? What are they doing? Why was this scene photographed? What happened next?

Get Active 1.4

Bring in a personal photograph. You can use the photograph that you used for the previous activity, but make sure you know all the relevant details, such as who is in the photo, when and where it was taken, what was the occasion, etc. Get into pairs and take turns to find out the story.

a) Your partner can ask you a maximum of five questions about the picture using the 'five Ws': who, why, where, when and what?

b) He/she then relates to you the story behind the photograph.

c) You tell your partner if he/she is correct and then provide any extra details to give the full story.

d) Change roles: it is your turn to look at your partner's photograph and do the same activities.

e) Discuss how you got on.

A forest ecosystem

Learning intentions

In this chapter, I am learning:

- how to read photographs to understand the component parts of an ecosystem
- what an ecosystem is and how it works
- how ecosystems are affected by people
- how to plan, create and manage a pond ecosystem.

What is an ecosystem?

An **ecosystem** is a living **community** of plants and animals and the environment (home) in which they live. The living things interact with their environment and with each other.

Components of an ecosystem

This photograph shows an ecosystem with some of the component parts named. All of these named parts are linked together because they interact and influence each other.

Ecosystems are all around you! They can be in parks, gardens, fields, forests, ponds, lakes, oceans and mountains. They can be any size or any scale – as large as a rainforest or as small as a goldfish pond. Your school grounds are an ecosystem, and the Earth's surface is an enormous ecosystem, so there is no common size.

Whatever its size, every ecosystem is made up of:

- living elements – these include plants, animals, humans and bacteria
- non-living elements – such as sunlight, air, water, rocks and soil (the environment).

Lough Neagh, Northern Ireland

Farmland ecosystem near
Camlough, County Armagh

Get Active 2.1

Look at the ecosystems shown in the photographs.
a) Take each photograph in turn and:
  • list all of the living parts in each ecosystem
  • list all of the non-living parts in each ecosystem.
b) Have you really listed everything, or only the most obvious things that you can
   see in the photographs? What can't you see in the photographs, but might be part
   of each ecosystem? Add these to the lists.
c) Compare your lists with a partner.

How does an ecosystem work?

Green plants contain a pigment called **chlorophyll**, which makes leaves green. Plants trap the sun's energy in chlorophyll to make their own food in a process called **photosynthesis**. The diagram below shows how this works.

3   The sun's energy turns the carbon dioxide and the water into **glucose**.

2   Leaves take in **carbon dioxide** and release oxygen back into the **atmosphere**.

1   During photosynthesis, water is absorbed by the roots of green plants and is carried to the leaves.

4   Glucose combines with minerals from the soil to make the plant grow. These minerals are called **nutrients**.

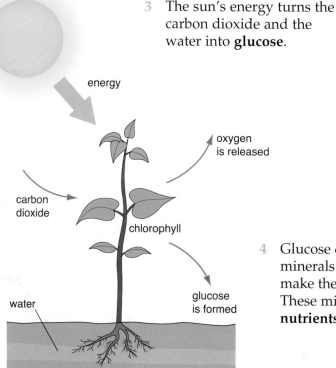

energy

oxygen is released

carbon dioxide

chlorophyll

water

glucose is formed

Photosynthesis

All animals in an ecosystem depend on plants for their food. Without plants, all living things will die. Here are two examples:

- A weasel eats a mouse that was feeding on clover in a field.
- A fox eats a skylark that has just eaten a spider, that had eaten a butterfly, that was feeding on nectar from plants.

So, in every ecosystem there are:

- **Producers** – Green plants are called producers because they can produce their own food.
- **Consumers** – Animals cannot make their own food. They are called consumers because they get their energy by either eating (consuming) plants or other animals.
- **Decomposers** – Fungi and bacteria in the soil feed on dead animals, droppings and other **waste** material. They make things rot, so they are called decomposers. These are the **recyclers** of the ecosystem. When rotting material is broken down, nutrients are released into the soil that can then be used by growing plants. Everything dies in an ecosystem but nothing is wasted and all is reused!

Food chains

The flow of energy from plants to animals is called a **food chain**. A food chain shows what eats what, and always starts with a plant. Energy trapped from the sun by the plant passes down the food chain. Energy is therefore essential to all ecosystems, because without a supply of energy, plants would die and life could not continue.

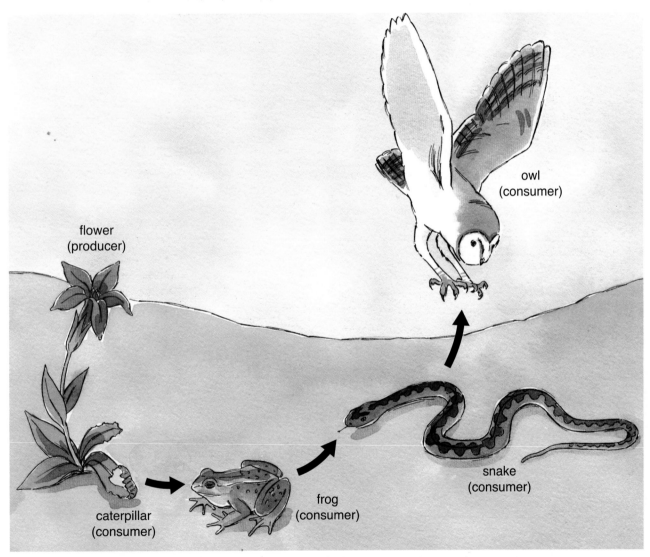

owl
(consumer)

flower
(producer)

snake
(consumer)

frog
(consumer)

caterpillar
(consumer)

Who eats what in a food chain

Animals that just eat plants are called **herbivores**. Animals that eat other animals are called **carnivores**. **Omnivores** are animals that eat both plants and other animals. What do you think humans are? Where would you put humans in a food chain?

Get Active 2.2

a) Arrange the following species into a simple food chain using arrows to show the flow of energy: frog, grasshopper, fox, plant material, spider.

b) Make up two simple food chains that include humans.

Odd one out

Work with a partner.

c) All the things in the table on the right are related to ecosystems. With your partner, find out the meaning of the words that you do not know (you can use Google™ for your research).

d) For each set of three words in the list on the notepad below, decide:
- Which word is the odd one out? Why?
- What connects the other two words? Why?

sunlight	sundew	butterfly
rabbit	bracken	bog cotton
earthworm	rocks	frog
pond snail	pike	crane fly
perch	oak	wind
butterwort	tufted duck	heather
pondweed	humidity	bladderwort
spider	alder	soil

1. sunlight   heather   humidity
2. rabbit   frog   butterfly
3. rocks   oak   alder
4. pike   perch   cranefly
5. sundew   bog cotton   butterwort

e) Find one word from the table above to add to each set of words in the list on the notepad, so that the odd one out remains the same, but now there are three words that are related in each set.

f) Add two more sets of words to the list (with an odd one out each time). Exchange your sets of words with your partner, and work out the odd one out for each set.

g) **Sort** out all of the words in the table above into groups. Give a brief description of each group.

Food webs

Most animals eat more than one type of food. When more than one consumer feeds on the same plant, food chains become linked. The flow of energy through an ecosystem is illustrated as a **food web** like the one below.

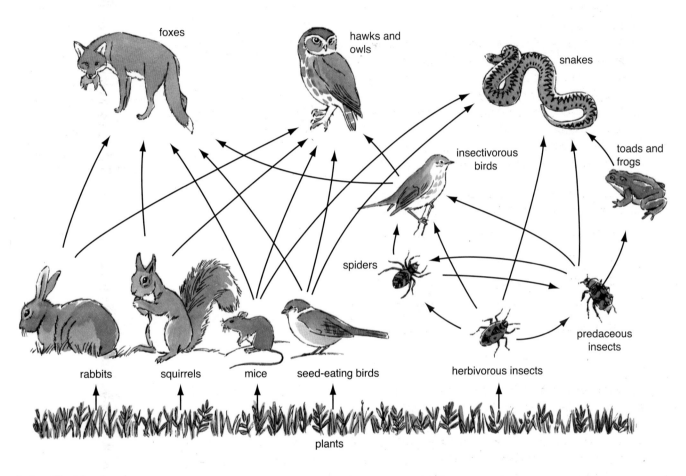

A simplified food web

A food web is where all the plants and the animals in an ecosystem are linked together, a bit like a spider's web. Energy is transferred by eating or being eaten! Each part is important, as each interacts and depends on the other parts. There is a natural balance.

If some event interferes with a food web, all of the other parts will be affected in some way. For example:

- If all the rabbits died: foxes, hawks and owls would have to eat more squirrels, mice and birds to survive; and the numbers of these animals would therefore decrease.
- If all the foxes died: the numbers of rabbits, squirrels, mice and seed-eating birds would increase; and so there would be more food for hawks and owls, so their number would increase too.

Get Active 2.3

Work in pairs.

Study the diagram of a food web (below) that could be found in a freshwater lake. What might happen if the events 1 to 3 (below) occurred a) in the short term, and b) in the long term?

Copy out the table below for your answers. You can make it much larger than this – perhaps use a whole A4 page to leave plenty of space for your answers.

Event	Short term (a few months)	Long term (after a few years)
1. The pondweed stopped growing.		
2. The bloodworms all died.		
3. The perch were removed.		

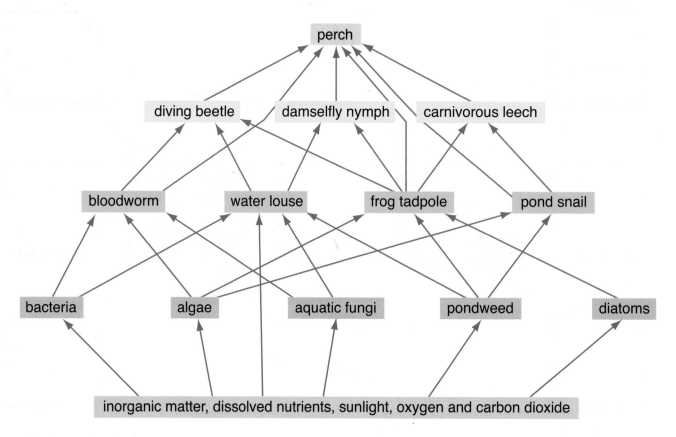

A freshwater food web

Types of ecosystems

Ecosystems can be found on land and on water. No matter where ecosystems are found on Earth, each will have living and non-living parts. There will be differences too, as different ecosystems work in different ways.

Get Active 2.4

a) Copy the diagram below and fill in the boxes to show some of the similarities and some of the differences between the two ecosystems shown in the photographs. The similarities will be the features that all ecosystems have in common. You can make the boxes bigger to leave more space for your answers.

b) Write five sentences to describe some of the things that you have learned about ecosystems so far.

Similarities and differences

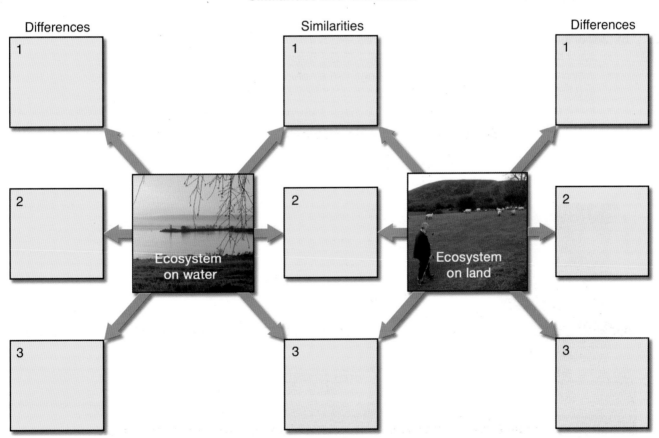

How we affect the environment

The Earth is one big ecosystem and humans play a powerful role in it. We make changes as we use many of the living and non-living parts in the environment around us. We use the Earth's **resources** to provide us with food, build houses and roads, keep warm, etc. We manage and control our environment to improve our standard of living. Unlike animals, humans make a big impact on the Earth's ecosystems. Sometimes we can damage ecosystems by destroying the environment in which certain kinds of plants and animals live.

Sustainable living means getting what we want from the Earth without wasting resources and destroying natural **habitats**. To make decisions about how to live sustainably, we need to understand how our actions affect the different parts of the ecosystems.

Get Active 2.5

Work by yourself or in pairs.
a) Look at the photographs below and on the following page.
- What human activity does each photograph show? Describe in detail.
- Why do you think humans made this change to the natural environment?
- In what ways do you think humans have benefited from this development/change?
- In what ways do you think humans have suffered from this development/change?
- In what ways have the ecosystems changed as a result of the development?
b) Be prepared to talk to the rest of the class and to your teacher about the ideas you have just come up with.

A building site outside Banbridge, County Down

Trying to satisfy the increasing demand for housing

The out of town Sprucefield shopping complex, Lisburn

A caravan site on sand dunes in Portrush, County Antrim

The endless journeys on the M1!

Coniferous woodland in Garvagh Forest, County Derry

Are humans dirty animals?

'Humans differ from all other animals in being a particularly dirty species. In contrast to other animals, humans are animals that foul their own nests and pollute their environment.'
Joy Tivy and Greg O' Hare

Humans damage ecosystems by polluting the environment. Producing vast amounts of waste and creating litter is one way in which we do this. Not only does it spoil the look of the environment, but it can have more serious **consequences**:

- Humans and animals can be injured by broken glass.
- Substances can leak into the ground and then into rivers and lakes, and poison fish and other waterlife.
- Seawater for bathing can be spoiled by **sewage** pouring into it.

Fly tipping in a river in County Tyrone

Dead pollan as a result of pollution in Lough Neagh

More fly tipping in County Tyrone, including a dead sheep

A polluted stream in Northern Ireland

Leachate from a landfill site

Get Active 2.6

Look at the photographs on this page and opposite showing environmental pollution in Northern Ireland.

a) Look carefully at each photograph. What is the story here, as you see it?

b) Imagine you are a reporter from a local newspaper or a local radio. You have visited one of these sites and want to inform your readers/listeners about:

- the dumping of waste
- the impact of such activities.

A reporter needs to be clear, interesting and informative. What are you going to say?

c) Prepare a short report for the newspaper/radio show. Be prepared to speak to your audience – the rest of your class!

What is the way of the future?

People who litter and pollute the environment often do not know about the impact of their actions.

- Do we want our local areas spoiled by such activity?
- Can something be done to bring about change?
- What strategies do you think can be used to stop humans causing pollution to the environment?

Get Active 2.7

The class divides into four groups:
- Two groups do Task A (representing scenario 1).
- Two groups do Task B (representing scenario 2).

a) Each poster must:
  - be at least A2 in size
  - show clearly what it is about without anyone having to ask any questions (it must hang alone and speak for itself)
  - have at least four colours, four images, a title, and accurate spelling and grammar
  - have at least one moving part.

b) Each person in the class is to judge the posters. You can award each poster a mark out of ten according to the **criteria** set out above. (You must be able to justify your marks, so briefly explain your marking system.)

c) Give your marks to your teacher, who will give the overall results and announce the winning poster!

Extension

Task A: Scenario 1

Design a poster with the title:
'50 Years of Humans, the Dirty Animals'.
Create a dramatic picture to show what could happen in 50 years if nothing is done to stop the dumping of waste. Use your imagination to do this! You could include photographs, newspaper headlines, pictures that you have drawn, a collage of waste/litter that you have found lying around, etc.

Task B: Scenario 2

Design a poster with the title:
'We want to stop it NOW!'.
Illustrate what can and must be done to ensure that there is less pollution in the world. You could include photographs, newspaper headlines, pictures that you have drawn, information you have collected, etc.

a) Research the damage caused in Northern Ireland by humans as they pollute the environment. The internet and newspapers are good sources of information. Remember you can use your own local area and conduct your own research. Useful websites include:
  - www.rspb.org.uk
  - www.ehsni.gov.uk
  - www.foe.co.uk.

b) Decide how you are going to present your findings to the rest of the class. You want to make an impression, you want people to listen and learn about the cause of the pollution and the effect. You want them to start asking questions, you want to make them think. Each group will have three or four minutes to present. A video or a PowerPoint® presentation could be very effective ways to illustrate your talk. You could download a soundtrack from Audio Network in the Net work area of LNI (or you could create your own soundtrack to create the right atmosphere).

The Big Task

Your task is to design an attractive area in the school grounds consisting of a pond ecosystem, with a seating area and with some waste bins.

The seating area and bins must be created out of **recycled** materials. This task will develop many useful skills for you and the members of your group. The task will be more easily achieved if you take time to plan properly.

You need to clarify the following issues:

- What is the big picture? What is your overall aim? What do you want to achieve?
- What are the steps that you need to take, what are all the smaller jobs that need to be done to achieve the overall result?
- Who will do what?
- Where will you research the information that you need?
- What will be the priority tasks (what needs to be done first, second, etc.)?
- Who will collect the information obtained?
- Who will sort the information and where will it be stored, so that it can be easily referenced and will not get lost?

There will need to be a lot of discussion within your group to work out an agreed plan of action. As you do this task, it will be important to record why you decided certain things. You must be able to justify all decisions that you make, so always write down why you chose particular species, designs, etc.

You can get help and advice to do this task from people who have the skills that you need.

The plan of action

1. Choose an area within the school grounds where it would be possible to have the pond.
   - Map the location. Try to work out the size of the pond.
2. Research the creation of a pond.
   - How do you make a pond? What is the ideal size? Which species of plants and animals are required to enable it to function (remember food chains and food webs)? When is the best time of the year to stock the pond? What maintenance is needed? How is it kept clean, and how do you deal with the possible problems that might happen?
3. Plan where your seating area is to be located.
   - What sort of seats do you need, and what recycled materials could you use to make them? Where could you get the materials? How do you make the seats? What would you make the bins from?
4. Create a design for the pond.
5. Choose the plants and animals to live in and feed from the pond.
   - You want to have an attractive pond with good, functioning food chains and food webs.
6. Draw a plan and picture of the pond and seating area as it will look when finished.
7. Create a wall presentation of all of your work. Include a paragraph stating clearly what your group feels would be the benefits of this project to the whole school.
8. Review and reflect on what you have done.

3 The changing world of work

Shoppers in Belfast town centre

Learning intentions

In this chapter, I am learning:
- to classify work types
- to read and represent statistics about change in employment structure
- about the ethnic minority groups working in Northern Ireland
- to think, problem solve and make decisions about changes in the world of work.

How do we earn a living?

All the people in the pictures above are working to earn a living. They are involved in **economic activity**, which means they get paid for their work. They are all doing different jobs in different places.

Get Active 3.1

Look at the photographs below showing three people at their work. Each pair of photographs show a different type of work – the farming, manufacturing and service industries.

a) What is each person doing? Describe what you can see about each person's work and the place in which they work from the photographs.

b) List the skills each person would need to do their work.

1A

1B

2A

2B

3A

3B

Types of economic activity

You may think that there are many types of jobs, but in fact, each job can be grouped into three categories: **primary**, **secondary** or **tertiary**.

Industry	Job	Economic activity
Primary	**Extracting** or using natural resources from the ground or sea.	**Agriculture**, fishing, forestry, mining, quarrying.
Secondary	Natural resources such as crops grown in farms, or **coal** mined from the ground, are **processed** for other uses or to make new products.	Manufacturing, **processing** or **assembling**. For example, rubber is tapped from trees and is used to make tyres in one factory. Then it is sent to a car manufacturing plant in another place for making cars.
Tertiary	Provide **services**, such as nurses, teachers, shop assistants and office workers.	Transport, **retail**, health, education, entertainment, banking, **tourism**.

Get Active 3.2

Odd one out

a) Look at all the different jobs in the table below. They can all be classified into a job of work in primary, secondary or tertiary industry. The list of sets of numbers in the box below relates to a job in the table above it. Write down which job is the odd one out in each set. Each time explain why this job is the odd one out, and what connects the other two numbers.

1. Teacher	6. Baker	11. Shipyard worker	16. Engineer in a quarry	21. Engineer in a forklift factory
2. Farmer	7. Machine operator in an aircraft design company	12. Post office worker	17. Oil rig worker	22. Fisherman
3. Furniture maker	8. Miner	13. Forester	18. Worker in a food processing plant	23. Doctor
4. Shop manager	9. Jewellery maker	14. Fitter in an engineering company	19. Secretary	24. Banker
5. Chef	10. Fireplace manufacturer	15. Horticulturalist	20. Politician	25. Dentist

b) Add an extra number to each set, but keep the same odd one out.

c) Put together your own group of numbers with an odd one out (make two sets of three numbers).

d) Sort out all of the numbers into three groups.

Set A	1	12	15
Set B	16	3	22
Set C	21	18	24
Set D	11	23	19

Extension

Work in pairs or small groups.

Use the job section from newspapers, such as the job finder section from the *Belfast Telegraph* and the jobs section from a local newspaper.
a) Make and cut out headings under which you can group different jobs: Primary, Secondary, Tertiary.
b) Cut out all the job adverts. **Classify** each job, putting it in a pile under the appropriate heading. If you find any jobs that are difficult to classify, put them in a pile on their own.
c) Count the number in each pile:
   • Which pile has the greatest number of advertisements for jobs?
   • Which pile has the least?
   • Which jobs were difficult to classify? What extra information would you need to help you to classify these jobs?
d) In pairs, talk about your results. Discuss reasons for your findings.

How is the world of work changing?

Statistics are useful for recording information about the changes in the world of work.

Year	Primary (farming, fishing, forestry)	Secondary (manufacturing)	Tertiary (services)
1973	25,440	166,580	250,370
1978	19,760	141,150	302,150
1983	17,920	105,190	330,240
1988	17,670	103,960	356,600
1998	16,060	106,680	453,640
2003	14,710	91,890	525,270

Numbers of people employed full-time in Northern Ireland in primary, secondary and tertiary industries during the years 1973 to 2003.
Source: Department of Enterprise, Trade and Investment (DETI)

Get Active 3.3

Work in pairs.
a) Look at the statistics in the table above. What **trends** do you notice in the 30-year period? Discuss what you think with your partner.
b) Each person makes a note of the trends identified.
c) Decide a method you could use to show one of the trends more clearly in a graph.
d) Suggest reasons to explain some of the trends that you identified.
e) Share your ideas with the rest of the class.

The changing pattern of economic activity

Life is very different today in Northern Ireland from how it was just 30 years ago!

- The number of people working in primary and secondary industries has declined.
- There has been a growth in the number working in the service industries. This situation is the same in all the **more economically developed countries (MEDCs)** in the world.
- The biggest drop in employment has been in **secondary industry**. There were more than 160,000 jobs in secondary industry at the beginning of the 1970s but now there are less than 92,000.
- Jobs in traditional manufacturing industries that Northern Ireland was famous for have now declined, such as: shipbuilding, heavy engineering, linen and other textiles.

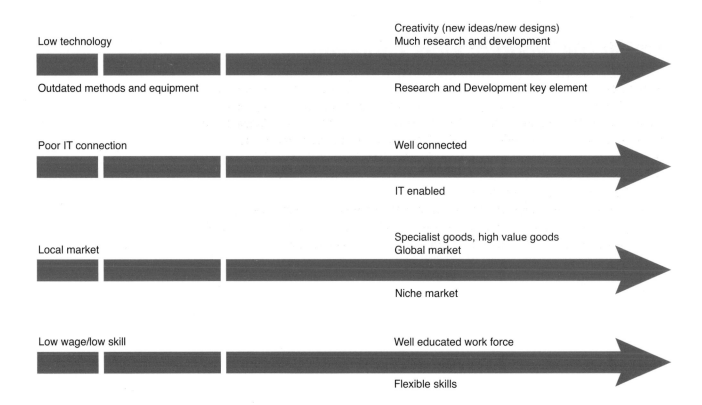

Looking towards the future of manufacturing in Northern Ireland. Source: adapted from PWC report

Get Active 3.4

Study the diagram on page 40 carefully and then answer these questions:

a) What is the purpose of the diagram? What is it showing?

b) Can you pick out four problems that exist in some of the present manufacturing industries? Write one sentence about each.

c) What needs to be done if Northern Ireland manufacturing firms are to have a successful future? Write a short paragraph (including four items of information from the diagram).

d) What do educational establishments (schools, further education colleges and universities) need to do to make sure that their students are well educated for modern industry?

e) When planning for your future, what do you need to do to ensure that you have the skills that employers want?

Extension

What's going on in Titanic Quarter?

Big changes are happening on the 185 acres that were once part of the famous Harland and Wolff shipyard. Shipbuilding here could not compete with other countries, and so it was forced to shut down. A decision was made to adapt to change in a different way. The area formerly used for this secondary industry is to be transformed into a new, mixed-use waterfront in Belfast. This is called **regeneration**. In this development there will be a science park, facilities for leisure and entertainment, hotels and restaurants, residential areas, preserved historic sites and more.

- Go to www.titanic-quarter.com.
- Explore the site. Now produce a colourful, illustrated information leaflet to promote this development. Your leaflet should explain where the development is (include a map) and describe what the land is to be used for.

The workforce

The rise of the service industry has led to a change in the **workforce**. Northern Ireland is increasingly becoming a **culturally diverse society**. This means that there are many people living and working in Northern Ireland, maybe as many as 40,000 people, who were born outside the United Kingdom.

Get Active 3.5

Work in pairs.

Read through the newspaper headlines below. Your task is to get the real story behind the headlines. Compile a list of questions about our changing workforce that you need answering in order to get the stories. Remember the 'five Ws':

| What? | Where? | Why? | When? | Who? |

STEP HOSTS MIGRANT SUPPORT CONFERENCE

Ethnic Minority Information evening in Kilkeel

Launch of new column for our Polish readers

RACIAL ATTACK VICTIM SUFFERED KNIFE WOUND

Young people complete education programme on anti-racism and promoting cultural awareness

Racist leaflets distributed all over town 'disgraceful'

Northern Ireland is becoming increasingly diverse and multi-cultural

Until the 1960s, the workforce in Northern Ireland was almost all English-speaking and white. People who have a different culture from the majority population in Northern Ireland are called **ethnic minorities**. The Chinese, Indian and Pakistani communities have lived here for the longest time and have come to regard Northern Ireland as home. Many have children and grandchildren who were born here. **Migrant workers** come to Northern Ireland to work temporarily before going back to their home countries. Industries have come to rely on migrant workers because:

● there are not enough people in Northern Ireland with the skills needed
● there are not enough people in Northern Ireland willing to do some particular jobs. (Which jobs do you think these are?)

Get Active 3.6

Look at the map of the world below. You will need to use an atlas for this task.

a) Identify the countries shown and write down the name of the country indicated by the numbered flag pole.

b) Work in groups. Each group needs to research some interesting information about one of these countries. You might choose to find out about the climate, food, language, traditions, or any other details that you think are interesting. Include

pictures and maps to illustrate. The internet and the library, and travel books/magazines will be useful sources of information.

c) Find out how to write down some words or phrases in the language of the country that you are researching.

d) When all of the information is collected, you can then make a colourful wall display of your work in the classroom.

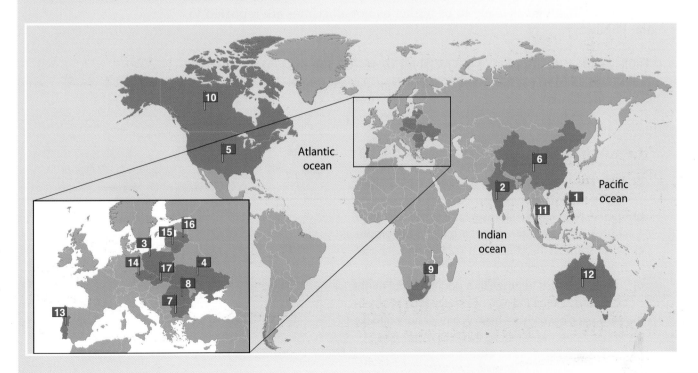

How well do immigrants integrate in Northern Ireland?

In the Northern Ireland Life and Times survey 2002, a sample of people in Northern Ireland were asked questions about **immigrants** in Northern Ireland.

- Only 33 per cent believed that immigrants were good for Northern Ireland's **economy**.
- Almost 50 per cent believed that immigrants take jobs away from people born in Northern Ireland.

These are **misperceptions** (mistaken beliefs). Migrants that have come to Northern Ireland provide skills that are needed by the country, often in industries where it has proved difficult to recruit workers locally. Immigrants have contributed much to the Northern Ireland economy.

Northern Ireland has recently been dubbed the 'Race Hate Capital of Europe'. This means that in Northern Ireland there is a problem of **racism**. This is often because people find it hard to adjust to a multiethnic society.

Look at the two tables on page 45. The information shows some of the problems of cultural integration as more migrants come into Northern Ireland to work. Table 1 shows the number of **racial incidents** reported to the PSNI. The incidents include: criminal damage, theft, robbery, intimidation/harassment, assault, threat or conspiracy to murder. One murder was committed in 2004/2005.

These statistics represent the racial incidents *reported* to the PSNI. The actual number of incidents is likely to be higher as not all crimes are reported to the PSNI.

Table 2 shows the location of the racial incidents (during the period 2004–2005 and 2005–2006) in each of the Local Policing Areas (referred to as District Command Units).

Get Active 3.7

a) Study tables 1 and 2 shown opposite. What trends do you notice?

b) Choose two ways to represent the statistics shown in Table 1 (for example, a line graph or a bar graph).

c) Select some of the statistics in Table 2 and choose a way to represent them.

d) Talk about these trends with a partner. What might explain each trend that you noted (see Prompt box 1)?

e) Many people who have been victims of **hate attacks** do not report the crimes to the PSNI. Why? (See Prompt box 2.)

f) Why is it wrong to base actions on misperceptions?

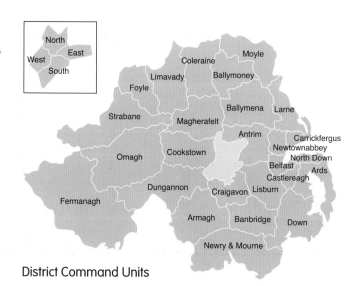

District Command Units

1998/99	99/2000	2000/01	2001/02	2002/03	2003/04	2004/05	2005/06
93	237	260	185	226	453	813	936

Table 1 The number of racial incidents reported to the PSNI

District Command Unit	Racial incidents 2004–05	Racial incidents 2005–06
Antrim	23	37
Ards	25	28
East Belfast	54	74
North Belfast	76	62
South Belfast	159	106
West Belfast	13	13
Carrickfergus	21	6
Castlereagh	12	21
Larne	0	2
Lisburn	28	43
Newtownabbey	15	52
North Down	19	20
Armagh	14	20
Ballymena	52	53
Ballymoney	8	7
Banbridge	1	4
Coleraine	22	37
Cookstown	16	21
Craigavon	79	78
Down	8	13
Dungannon & South Tyrone	69	102
Fermanagh	18	29
Foyle	40	27
Limavady	7	7
Magherafelt	9	17
Moyle	1	7
Newry & Mourne	15	35
Omagh	6	13
Strabane	3	2
TOTAL	813	936

Table 2 The location of racial incidents during 2004–2005 and 2005–2006

Prompt box 1

These statements could help you answer question d). Remember that some of these are mistaken beliefs (misperceptions) about the immigrant population.
● They look and sound different.
● Some people say they don't mix.
● They take our jobs.
● They take over our areas as there are so many of them now.
● We don't like them.
● They are easy targets and don't fight back.

Prompt box 2

These statements could help you answer question e):
● Cannot speak good English.
● Afraid to make a fuss in case it makes things worse.
● Have been told by some in the community not to have anything to do with police.
● Lack of confidence in police.
● Bad experience of police in own country and believe PSNI would be the same.
● Unsure of rights.

A case study

In many social situations, people adopt different viewpoints. In other words, there are many ways of looking at things. Every point of view may seem right for the person holding it, but not right enough to be imposed on others. This task is called OPV (other people's views). It is designed to help you collect the different views of all the people affected by a situation. Sometimes when we listen or even read about the views of other people, it can be possible to escape our own, often narrow, point of view, and so take other people and other views into consideration.

This story is from a PSNI report.

1. A Chinese family moves into your area. This is the only Chinese family in your street. They feel as if they are not welcome – no one speaks to them and neighbours appear to cross the road. People don't seem friendly. Perhaps it is because they are new.

2. The local social club has no room for new members and the family appears to be served last at the local shops and café. They have bought their child a cat and it comes home one evening sprayed with paint. Their child starts at a local primary school and comes home the first evening crying; the other children were calling her names. The car gets scratched; the front room window gets smashed.

3. They hear banging at the door at night and when they go to look no one is there. All sorts of things are thrown into their garden. Constant looks and mutterings from passers-by 'we don't want you here', 'why don't you go back home', or are they imagining things?

Thankfully this family is OK. If it weren't for the smoke alarm things could have been much worse.

4. The graffiti begins first on the fence and then on the door. The car tyres are slashed and then the firework comes through the letterbox one winter evening and sets the curtains alight.

5. This is the first time the police have been contacted.

Get Active 3.8

Divide into groups of five. Elect a leader/manager in each group who will organise the activity.

Read the account above of a Chinese family's experience in Northern Ireland.

a) Each person in the group 'becomes' one of the persons affected by this incident: the mother, the father, the child at the primary school, an unfriendly neighbour, and the person who has been harassing/attacking the family. Each person should try to imagine what the person they are portraying thinks and feels.

b) Write an account of what has been happening to you (from your chosen character's point of view). Include:
* what has been going on
* why you think it has happened
* how you have behaved
* how you feel about it now.

c) Each group presents their set of views to the rest of the class. After listening to all the viewpoints, discuss together with your teacher:
* Is it easy to see other people's viewpoints?
* If other people cannot see your point of view, should you bother with them?
* Why is it necessary to see someone else's viewpoint?
* Should you think about other people's views before you take action about something you feel strongly about?

A vision for Northern Ireland?

This statement was made recently by the Equality Minister, Jeff Rooker.

*'Our vision for Northern Ireland is of a **society** in which racial diversity is supported, understood, valued and respected: a society where racism in any of its forms is not tolerated and where we live together as a society and enjoy equality of opportunity and equal protection.'*

This is the vision for the future; for a multicultural Northern Ireland.

The Big Task

You are part of an organisation that helps new migrants settle into Northern Ireland.

1. What are the important things that your organisation needs to do? Write down everything that you can think of that is:
   - necessary
   - useful
   - neighbourly.

The information box on the right identifies some of the main problems that migrants to Northern Ireland have experienced.

2. Read all of your suggested actions through. Think carefully about the changes that they could bring about in the lives of people that have recently migrated to Northern Ireland for work. It isn't always possible to put all suggestions into action, no matter how good they are. In many cases it may be necessary to select the suggestions that would be of greatest value or the actions that are most urgent.

3. Now select your 'Priority actions'.
   - Place an 'x' beside the five most important actions in your opinion.
   - Be prepared to talk about the reasons why you think that the five suggestions you have chosen would make a difference.

Information box

When interviewed, many of the migrants to Northern Ireland identified these problems.

Lack of information on:
- home security
- personal safety
- how to buy a car, tax and insure it
- how to get work and undertake training courses
- rights at work (holiday entitlement, sickness and maternity benefit).

Difficulties:
- problems with the English language
- racism and harassment
- underpayment at work and late payment for work done
- some qualifications aren't recognised here
- loneliness and isolation.

4 Northern Ireland rocks!

The rocky coastline of County Antrim

Learning intentions

In this chapter, I am learning:

- to identify the main rock types found in Northern Ireland
- how the three main types of rock are formed
- how rocks are important for people.

What is Northern Ireland made of?

White Park Bay in County Antrim

Rocks form the solid crust of the Earth. In some places in Northern Ireland you would have to dig down through the soil just a few centimetres to get to solid rock; in other places you would have to dig for a metre or more, but everywhere you would come to solid rock.

There are two sorts of solid rock in this photograph of White Park Bay in County Antrim. Sticking through the sand of the beach is a white rock while the cliffs in the background are black. But, as you will find out, they have many more differences too.

People who study rocks are called **geologists**. The first thing to know about geologists is that they think of time very differently. You may think that 100 years is a very long time, but for a geologist that is hardly any time at all! This is because rocks take a long time to form. A rock that is 10 million years old is a young rock to a geologist!

Get Active 4.1

a) What names of rocks do you know? Brainstorm a list. Then compare your list with the people sitting around you, and add to it.

b) Do you know the most common rock near to where you live? Where would you go to see this rock?

c) Work in groups. Produce a 30-second advertisement for radio to encourage young people to become geologists. Your team must first research the job of a geologist: what they do, where they work, what qualifications are needed, and so on.

What kinds of rocks are there?

Every kind of rock on the Earth can be sorted into one of three groups, according to how it was formed. These groups are **sedimentary**, **igneous** and **metamorphic**.

Sedimentary rocks

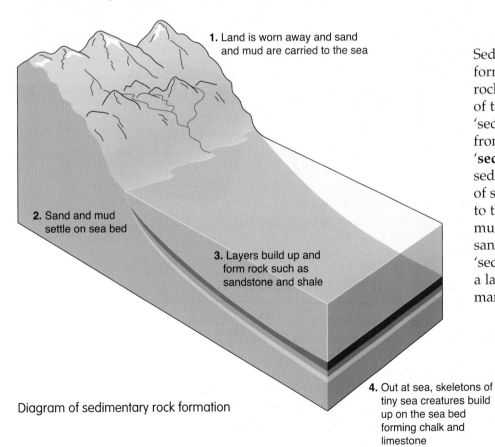

1. Land is worn away and sand and mud are carried to the sea

2. Sand and mud settle on sea bed

3. Layers build up and form rock such as sandstone and shale

4. Out at sea, skeletons of tiny sea creatures build up on the sea bed forming chalk and limestone

Diagram of sedimentary rock formation

Sedimentary rocks are formed from bits of older rocks, or from the remains of tiny creatures. The name 'sedimentary rock' comes from the same word as '**sediment**'. The word sediment describes the bits of sand and mud that drop to the bottom of a glass of muddy water. Mud and sand also settle as 'sediment' at the bottom of a lake or the sea over many, many years.

Get Active 4.2

a) Copy and fill in the blanks in the statements below. Sketch an appropriate picture to represent each statement.

| Earth's | sea | buried | rock | worn | carried | layers |

How sedimentary rocks form

- Rocks on the _____ surface are broken up and _____ down by **erosion**.
- Eroded material is _____ to the sea by **transportation** in rivers.
- Pieces of rocks settle on the _____ bed as **sediment**. This process is called **deposition**.
- Over millions of years _____ build up. Eventually the lowest **layers** are turned into _____ by the weight of the layers on top.
- When a creature is _____ in the layers it is turned into a **fossil**.

b) Write a definition in your own words for each word in **bold** in the statements above.

Types of sedimentary rocks

As you have just seen, sedimentary rocks take millions of years to form. Here are some of the most popular kinds of sedimentary rocks.

Sandstone forms when sedimentary rock is made from sand.

Coal is formed from dead plants.

Mudstone is made from mud. Sometimes it's also called **shale**. This sample has a fossil.

Chalk, shown here, and **limestone** are made from the skeletons of tiny sea creatures.

Two sedimentary rocks found in Northern Ireland

Shale

This shale was collected in Donaghadee on the Ards Peninsula and was laid down as mud about 430 million years ago. Shale has much smaller grains (up to 0.05 mm) than sand grains (which are up to 2 mm in size). The grains in shale are too small to be seen without a magnifying glass.

Chalk

The chalk that makes up the White Rocks at Portrush and at White Park Bay as shown here was **deposited** about 100 million years ago. Under a powerful microscope, chalk can be seen to be made of tiny skeletons of creatures and plants that once floated in the seawater. When they died they fell to the bottom of the sea as a mushy layer, and over time turned into chalk. Limestone is similar to chalk, but is less pure. This means it contains other kinds of sediment as well as skeletons of creatures.

White Park Bay (below) and the White Rocks at Portrush (right). The black rock above the chalk (in the photograph on the right) is an igneous rock (see pages 54–7). Which type of igneous rock do you think it is?

Get Active 4.3

a) How many of these tiny 0.05 mm grains from shale could you fit end-to-end on a 30 cm ruler? Let's try to work it out. 30 cm is 300 mm and you could get 20 of the 0.05 mm particles on 1.0 mm so that makes …?

b) How does this compare to sand grains?

c) Find out about some of the other sedimentary rocks found in Northern Ireland. (You could use the internet and a search engine, such as Ask™.)

For example, here are the names and photographs of three sedimentary rocks found in Northern Ireland. We put in the key words 'Geology Northern Ireland **Greywache**' to find the first one.

d) Find out more about chalk and limestone.
- How are they different?
- What are they used for?
- Where would you go to find them?

Greywache

Limestone

Extension

On the Ordnance Survey map (page 136) find the White Rocks near to Portrush.

What sort of rock does the geology map on page 60 show there? It is not chalk, but we know the White Rocks at Portrush are made of chalk. Puzzled?

a) Use the photograph on page 53, taken near to Portrush, to work out what is happening.
- What do you think has happened?
- What does the photograph tell us about the ages of the rocks? Which is older?
- Can you name the two types of rocks shown?

b) Work with a partner to put together a poster or a PowerPoint® presentation on sedimentary rocks.

Sandstone

Igneous rocks

The name of igneous rocks comes from the same word as 'ignite' which means 'to do with fire'. This is a good name for these rocks because igneous rocks form when a rock gets so hot that it melts. This is called **molten** rock (or magma).

There are two main igneous rocks in Northern Ireland: **basalt** and **granite**. They look very different and were formed from molten rock in very different ways.

This photograph shows **lava** pouring out of Kilauea Volcano in Hawaii

Basalt

Basalt is a very dark, almost black, rock. It was formed when molten rock cooled very quickly, such as when volcanic **lava** reaches the surface after an eruption. You will notice that the crystals in basalt are difficult to see without a magnifying glass. When a rock is formed on or near the surface of the Earth it cools so quickly that crystals do not get a chance to grow. Basalt can be used to surface roads, so there are a lot of quarries in basalt areas. The photographs below show some places in Northern Ireland which have basalt rocks.

The Giant's Causeway is the most famous volcanic rock formation in the world because of its hexagonal-shaped columns. The Antrim Plateau is a lava plateau made of basalt that covers most of County Antrim.

Granite

This rock looks about as different as it is possible to be from basalt. However, it is an igneous rock as well, made when molten rock cooled down. This time it cooled deep below the Earth's surface, so it was much longer before the molten rock became solid. Because it cooled much more slowly, the crystals had a chance to grow.

Giant's Causeway

Antrim Plateau

Slemish Mountain is a stump of a volcano

The basalt here was produced about 65 million years ago. It is considered a very young rock to a geologist!

Uses of granite in Northern Ireland

Granite is a very beautiful rock if it is polished. It can be used in thin sheets for the outside of shops and offices and for expensive kitchen worktops. It is also good for gravestones because it is hardwearing.

Kerbstones and cobblestones in Belfast used to be made of granite from the Mourne Mountains – often the kerbs still have the stonemason's marks on them like the triangle here.

Get Active 4.4

a) Find a picture from the internet of a piece of granite rock clearly showing the three kinds of crystals. Sketch this and label each type of crystal:
- quartz – colourless glassy crystals
- mica – soft black flecks in the granite
- feldspar – a pink or white coloured crystal.

b) Working in small groups, do some research about igneous rocks. You should include granite and basalt and others, too. Look for some photographs of them and find out how they were made and what they are used for.

c) As a class, agree on criteria that would make a good poster about igneous rocks. In your groups make a poster showing igneous rocks, with a picture or a drawing of a volcano in the middle. Then when you have finished, use the criteria you have agreed to judge your poster and the posters of other groups.

Much of the Mourne Mountains in Northern Ireland are made of granite

Metamorphic rocks

The name metamorphic comes from two words: 'meta' means change and 'morph' means shape. It is the third type of rock and is different to sedimentary rocks that were formed from layers of sediment, and igneous rocks that were formed from molten rock cooling down. Metamorphic is the name given to a rock that has been changed to become a different type of rock due to the effect of great heat, enormous pressure or both. Mostly this takes place under the crust of the Earth near earthquake zones, where earth movement causes pressure and heat that changes existing rock.

In this photograph (taken in White Park Bay in County Antrim) the rock on the left was red-hot molten rock when it came up through the chalk rock on the right. It cooled down to make basalt. The band of rock in the middle has been changed from chalk, a sedimentary rock, into a metamorphic rock called **marble**.

White Park Bay, County Antrim

Like other metamorphic rocks, marble does not contain fossils, as the great heat and pressure destroys them.

Marble can be found in a lot of colours from pure white to black. One famous source of marble is from the west of Ireland. Connemara marble is mainly green.

Like limestone, marble can dissolve in rainwater. In dry conditions, such as in the Mediterranean, ancient buildings made of marble can survive.

Metamorphic rocks in Northern Ireland

Slate

This **slate** is a hard metamorphic rock that is easily broken into sheets. Slate was once quarried in some parts of County Down, but most of the slate we are familiar with in Northern Ireland is imported from North Wales. Slate is sometimes used for roofs and can be seen on many older buildings such as these mill cottages close to the Kellswater River in County Antrim.

Mill cottages close to Kellswater River, County Antrim

Schist

When heat and pressure change rocks, they become unrecognisable compared to what they were like before. **Schist** is a rock that has been changed in this way. Almost any rock can become schist if it is changed enough. In this picture it glistens with the mineral mica (also found in granite) so this is called mica schist.

Get Active 4.5

a) There are many other types of metamorphic rocks in Northern Ireland, such as **quartzite**. Find out about this rock and two others.

b) Draw three **concept maps** with the word 'GEOLOGY' at the centre and the words 'SEDIMENTARY', 'METAMORPHIC' and 'IGNEOUS' linked out from it. Put more links out and words around them and add more links to those words where they are needed.

c) Using the photographs of rocks in the chapter so far, and any more that you may have found in your research, make up a 'Who Wants to be a Millionaire?' game using PowerPoint®. Show the photograph and give four possible answers. Get the answer correct and you can go on to the next question! You can put in a 'Phone a Friend' (ask one person in the class) or 'Ask the Audience' (get the class to vote).

The Big Task

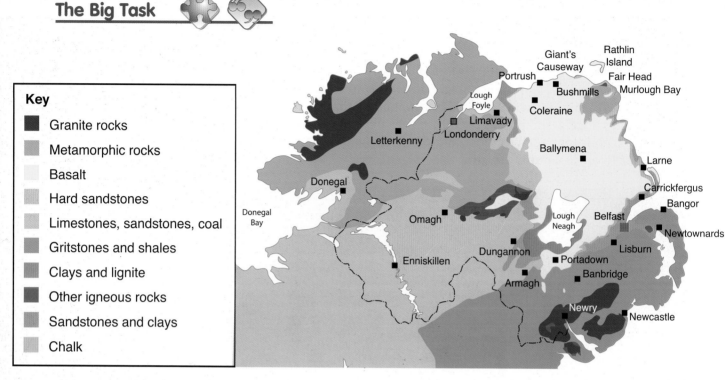

Key

- Granite rocks
- Metamorphic rocks
- Basalt
- Hard sandstones
- Limestones, sandstones, coal
- Gritstones and shales
- Clays and lignite
- Other igneous rocks
- Sandstones and clays
- Chalk

A simple geology map of Northern Ireland

1. Find where you live. What sort of rock does the map say is found there?
2. If you find a different rock in the soil in your garden or in a wall, does that prove the map wrong?
3. Make up three questions about the geology map and share them with a partner. Agree on your four favourite questions and ask another pair to answer them. The questions could be like the ones below or they could be quite different:
   - What rock are the Sperrin Mountains made from?
   - What is the rock type around Ballymena?
   - Why would south of Enniskillen be a good place for a cement works?
4. Work in a small group. How are one of each of the kinds of rocks used: sedimentary, metamorphic, igneous rocks (find out as many ways as possible)? For example, think about building, farming and making things.
5. Imagine that a quarry was planned to open in your area. What might the following people say about it?
   - Unemployed woman
   - Farmer
   - Factory owner
   - Politician
   - Environmentalist
   - Local resident

   What would you think? Why?

5 Rain, rain, go to Spain

A rainy, windswept coastal road in County Antrim

Learning intentions

In this chapter, I am learning:
- the difference between weather and climate
- to produce and interpret climate graphs
- to describe local weather and how weather varies
- how weather affects the lives of people
- to read and interpret weather maps and charts as well as satellite images
- how to conduct an investigation to observe, measure and record local weather conditions.

Is our weather really so bad?

The **weather** is a favourite topic of conversation for people. It is never far from our minds. What will it be like tomorrow? Will it be warm enough to go to the beach next time we are at the seaside? What will I wear tomorrow? Will we have a white Christmas? It is rarely out of the news.

Snow shuts down schools and roads

MORE GALES ON THE WAY

Crew rescued as storms batter coast

Big Freeze strikes Northern Ireland

FLOODS CAUSE CHAOS!

Sudden snow blizzards hit

High winds cause chaos in Northern Ireland

What is weather?

Weather relates to the state of the bottom 10 kilometres of the atmosphere. Weather describes the condition of the atmosphere at one place at one time. It refers to the day-to-day conditions: the temperature, the cloud cover, whether it will be wet or dry, the strength and direction of the wind, and so on. Weather conditions can vary, even over small areas. For example, the weather in Belfast could be very different from that in Derry, which in turn, may also be different from that in Newry.

Describing our weather

Much has been written about Northern Ireland's weather. In primary school, you may have come across different sayings about the weather:

Red sky at night, shepherd's delight; red sky in the morning, shepherd's warning.

Rain before seven, fine by eleven.

Clear moon, frost soon.

When smoke descends, good weather ends.

You can find more interesting sayings about the weather at www.metoffice.com/education/primary/students/sayings.html.

Get Active 5.1

a) Why do you think people made up such sayings about the weather?

b) From your experience, would you say there is any truth in such sayings about the weather?

c) All of us have experienced many different weather conditions in our lives, when living here in Northern Ireland or while abroad on holidays.
   - List all the words you can to describe the weather.
   - From the list, highlight those words that particularly describe Northern Ireland's weather.

d) Write a short poem or rap to describe the weather associated with any of our four seasons.

Our changeable weather: four seasons in one day!

Unlike some places in the world, our weather is not very predictable. The best word to describe Northern Ireland's weather is **changeable**. It can change quickly and often dramatically. In summer, a warm, dry day may be followed by a cool and wet day. You have probably heard it said that 'we can experience the four seasons in one day!'

Get Active 5.2

a) For each element of the weather listed in the table below, give the units of measurement and the name of the instrument used to measure it.

b) Draw a simple sketch of each of the measuring instruments you listed in the chart. Find out how each piece of equipment works. To find out more about this use this weblink to help you: www.bbc.co.uk/schools/gcsebitesize/geography/weather/mappingmeasuringrev6.shtml

c) Observe and measure the weather conditions in your school grounds to build up a picture of the general weather conditions. Your school may have a data logger connected to a PC which takes and records a number of weather readings at pre-determined intervals on a continuous basis. Alternatively, you may use more traditional measuring equipment.

Weather element	Units of measurement	Measuring instrument
Temperature		
Rainfall		
Wind speed		
Wind direction		
Cloud cover		
Pressure		

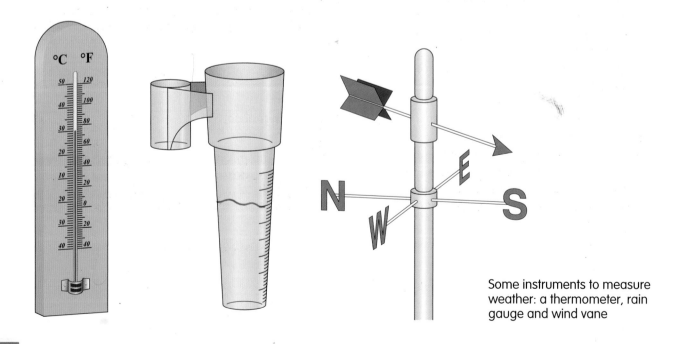

Some instruments to measure weather: a thermometer, rain gauge and wind vane

What is climate?

Climate is the average weather experienced by a particular area. It is a measure of what the weather in any area is normally like, based on measurements taken over a period of at least 30 years.

	Jan	Feb	Mar	Apr	May	Jun	Jul	Aug	Sep	Oct	Nov	Dec
Temp (°C)	8	7	8	13	13	18	19	19	18	14	9	7
Rainfall (mm)	81	64	46	91	94	23	46	78	19	84	93	54

The climate (average monthly temperature and rainfall) of Loughgall in County Armagh

Get Active 5.3

a) Use the data in the table to produce a **climate graph** for Loughgall. Look at the example of how to draw a climate graph on the right to help you.

b) Use your completed graph to answer the following questions about Loughgall's climate.
- Which was the wettest month? How much rain fell in this month?
- Which was the driest month? How much rain fell in this month?
- What was the total annual rainfall for Loughgall?
- Which was the hottest month? What was its temperature?
- Which was the coldest month? What was its temperature?
- What was the temperature **range** for Loughgall?
- Describe the annual temperature variations.
- Describe the annual rainfall pattern.

c) The climate of Northern Ireland is described as being **temperate**. What kind of weather conditions across the year characterise a temperate climate?

How to … draw a climate graph

a) Temperature and rainfall can be shown on the same graph. Temperature is shown on a line graph as a red line, and rainfall is shown by the blue columns on a bar graph.

b) The temperature scale is marked on the left-hand side in °C. Temperatures are plotted in the centre of the column for each month and the points are joined with a continuous red line.

c) The rainfall scale is marked on the right-hand side in mm (rain is called **precipitation**). The rain is marked for each month by a line – the columns touch each other and are shaded in blue.

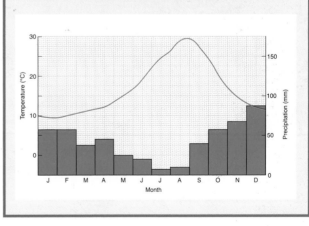

Some factors that affect our climate

Latitude: The further one moves from the Equator the colder it gets. This is to do with the curvature of the Earth and the angle at which the sun's rays hit it.

Distance from the sea: The sea warms up and cools down at a slower rate than the land. As a result, a sea breeze keeps coastal areas cool in summer, but warm in winter. The further inland you go the warmer it gets in summer, but the colder it gets in winter.

Prevailing wind direction: This refers to the most commonly experienced wind direction in a place. In Northern Ireland, the prevailing wind is from the southwest. This brings us a lot of rain from the Atlantic Ocean.

Altitude: Temperature drops as height above sea level increases. It decreases by 1.5 degrees Celsius for every 100 metres that you rise above sea level.

Get Active 5.4

a) Look at the climate data below for the Silent Valley in County Down. The Silent Valley is located in the Mourne Mountains.

b) Produce a climate graph for the Silent Valley to represent the data shown in the table. You can have a go at doing this using Microsoft Excel® (your teacher can give you guidelines for this).

c) Describe how the climate for the Silent Valley differs from that of Loughgall that you looked at earlier. Your answer should include reference to similarities, differences, maximum, minimum, range, average, seasons, and should include actual figures.

d) Look at the **physical map** showing the physical geography of Northern Ireland.

- Why is it always colder in Portrush in summer than in Malaga (southern Spain)? *Clue: look at a map of Europe in your atlas.*

- If Northern Ireland was to have a southeasterly prevailing wind, how would this affect our climate? *Clue: how do the geographical features to the southwest and southeast of Northern Ireland differ?*

	Jan	Feb	Mar	Apr	May	Jun	Jul	Aug	Sep	Oct	Nov	Dec
Temp (°C)	7	7	6	10	11	16	17	17	16	13	8	6
Rainfall (mm)	213	111	105	195	147	62	86	96	57	187	165	100

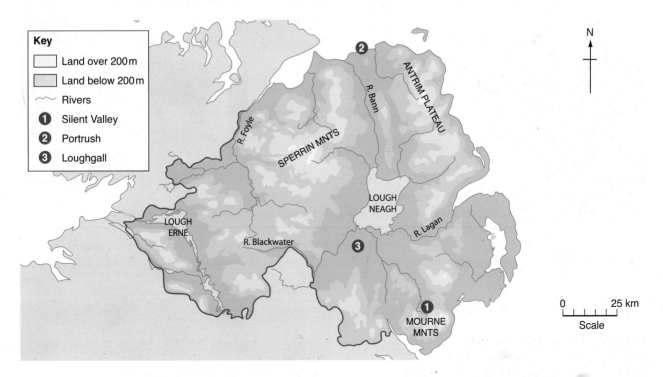

A physical map of Northern Ireland

How does the weather affect you and other people?

The weather can affect our lives in many ways, such as the clothes we wear, our leisure activities, the activities we do on holiday and so on. It may affect the jobs of other people and even make everyday life very difficult.

Get Active 5.5

a) Explain how the weather affects people in the eight photographs below and opposite.

b) Work in pairs. Using the 'five Ws' (What? Where? Who? When? Why?), write questions that will confirm whether what you think is going on in the photograph is actually the case.

c) Each pair shares their questions with the rest of the class.

d) The class agrees which are the most useful questions to use for the activity.

e) Each pair is given a copy of one of the photographs and an A3 sheet of paper.

f) The photograph is stuck in the middle of the page and the headings: 'What? Where? Who? When? Why?' are written around it.

g) Each pair discusses the questions and records their answers under the appropriate heading on the A3 sheet.

h) Each pair feeds back to the class what they have learned about how weather may affect the lives of people.

Flooding in the city

Flooding in the countryside

Feeding livestock

Getting about in the countryside

Making essential deliveries

Coastal storm

Trees block roads

Getting about in town

Is our weather here in Northern Ireland really as bad as we think it is?
It's time to dig a bit deeper to get a better picture for ourselves!

Get Active 5.6

a) Check out the following websites. Record the weather extremes in a table (similar
to the one below) for Northern Ireland, the United Kingdom and the world.
- www.meto.gov.uk (weather and climate > past weather > United Kingdom > Northern Ireland)
- www.meto.gov.uk/climate/uk/ (extreme weather)
- www.dandantheweatherman.com/Pikanto/Worldrec.htm

b) What have you learned in this activity?

c) What surprised you most/least?

d) How does the climate of Northern Ireland compare to the United Kingdom and the world?

e) Is our weather really so bad?

	Northern Ireland	United Kingdom	World
Highest temperature			
Lowest temperature			
Highest rainfall in 24 hours			
Highest monthly sunshine			
Lowest monthly sunshine			

The weather forecast

You can get a weather **forecast** for Northern Ireland from many sources: television, radio, newspaper and the internet. Below are various different weather forecasts for the same day in Belfast.

Today 2ⁿᵈ January 2007

7°		Light Rain Shower	High 7°	Humidity:	87%
11:50 GMT Tue			Low 6°	Wind:	W/26 km/h
				Visibility:	9.99 km
				Dewpoint:	5°
				Barometer:	Unknown
				Sunrise:	8:46
				Sunset	16:09

Source: www.yahoo.com

The weather for Northern Ireland on 2 January 2007

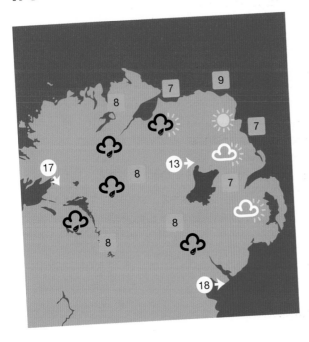

Source: BBC

Headline:
Showers at first, then mainly dry and bright. Cold. Windy.

Today:
Still some showers at first with snow on high ground. Showers soon dying out then dry and bright with sunny spells developing. It'll feel cold in the strong north westerly wind. Maximum temperature 7 °C.

Tonight:
Cloudy and becoming much milder overnight with outbreaks of rain. The rain will become heavy at times in north west but light and patchy elsewhere. Strong to gale force south westerly winds. Minimum temperature 5 °C.

Source: BBC

Northern Ireland Weather 02 January 2007

Latest observations: UK – 1100 on 2 Jan 07						
Location	**Weather**	**Temp**	**Wind**		**Vis**	**Pressure/trend**
			Dir	**Speed**		
Belfast (Aldergrove)		7.4 °C	WNW	14 mph	11 km	1022/Rising

Source: www.metoffice.gov.uk

Get Active 5.7

a) Look at the weather forecasts above and opposite. In what ways are the various weather forecasts similar and different?

b) Which elements of the weather are mentioned in all of the weather forecasts?

c) What symbols are used to represent the different weather conditions being predicted?

d) Which weather forecast do you like best? Why is this?

e) Which weather forecast is the most difficult to understand and of least interest to you? Why is this?

How accurate is the weather forecast?

Lots of people depend on the daily weather forecast and many need to be confident that the weather forecast is accurate. Can you think who these people might be?

Get Active 5.8

a) Make a list of three people who depend on the weather forecast and say why it is so important to them.

b) Look at the weather forecast for Northern Ireland. (You may choose to use television, radio, newspaper or the internet.) Create a simple table to record what the weather forecast predicts. Include:

- the date and details of temperature (maximum and minimum)
- precipitation (rain, hail, sleet or snow)
- wind speed and direction
- amount of cloud cover and any other relevant details.

This can be completed as a homework activity beforehand.

c) Describe the actual weather observed at your school on the day in question.

d) As a class, discuss how the weather forecast compared to reality:

- Which parts, if any, of the forecast were correct?
- Was the weather forecast completely wrong about anything that it predicted?
- Are some weather forecasts more accurate than others?
- What does this tell you about the nature of weather forecasts?

e) What might be some of the consequences of not having accurate weather forecasts? (What happens when the weather forecaster gets the forecast wrong?)

The Met Office and new technologies

Today our weather forecasts are fairly accurate. Weather forecasters in the **Meteorological Office** examine **satellite images** like the one below (pictures taken from satellites that continually orbit the Earth), to track the movement of **air masses** and **fronts**.

This was the satellite image for Europe at midnight on 2 January 2007

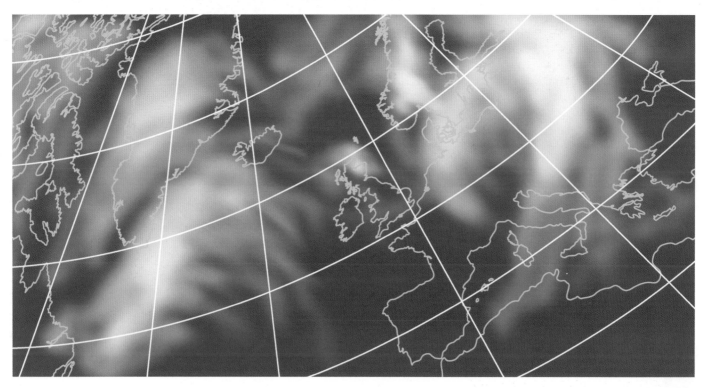

Source: The Met Office

These satellite images are used with weather recordings taken at numerous weather stations on land, and also by very sensitive equipment carried on aeroplanes, weather balloons and ships. The information is used to draw **synoptic charts** (like the one on page 73), which show the weather conditions at a given time. Increasingly, computers are doing most of this work.

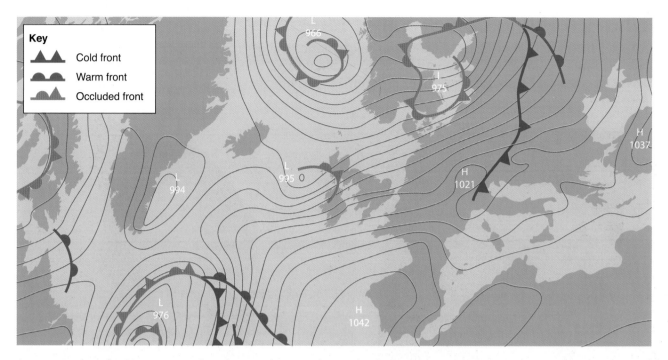

A synoptic chart for Europe, recording the weather timed at midnight on 2 January 2007

Get Active 5.9

a) On a copy of the satellite image on page 72 write the following labels:
1. A line of cloud stretching from the tip of North Africa, through Eastern Europe to Scandinavian countries.
2. A large band of cloud in the Atlantic Ocean.
3. Thick cloud over the Alps.
4. Clearer skies over Spain and Portugal.

Use an atlas to help you if necessary.

b) Study the satellite image and the synoptic chart above and state whether these statements are **true** or **false**.
1. Much of the British Isles and Ireland is covered with thick cloud.
2. It is likely to be raining in northern Scotland.
3. There is no cloud over the Atlantic Ocean.
4. Where there are no clouds – it must be warm and sunny.
5. To the west of Ireland is a ridge of high pressure.
6. This ridge of high pressure will bring a period of drier weather.
7. Further to the west is the next depression.

Extension

a) Find today's satellite image and synoptic chart for Europe:
www.metoffice.gov.uk > weather > Europe weather > Latest > satellite.
www.metoffice.gov.uk > weather > Europe weather > Forecast > pressure charts.

b) Print out a copy of both. Describe any similarities or differences between today's weather and that shown in the satellite image and synoptic chart here.

What's it like to be a weather forecaster?

Weather forecasters have become television celebrities and some have even had hit pop songs written about them. How would you like to be a weather forecaster? Could it be a career for you?

Frank Mitchell, ITV weather reporter

Angie Philipps, BBC weather reporter

Get Active 5.10

Work in groups.

a) You will have to produce a 30-second television weather forecast for Northern Ireland. To do this you will need to:
- watch and listen to television weather forecasts
- discuss how the weather forecaster goes about presenting the weather forecast (using maps, places, symbols, language)
- research what the weather is going to be like in Northern Ireland in the coming few days
- make a **storyboard** for your presentation
- write a simple script for the weather forecast
- include a suitable map with place names and weather symbols
- agree the criteria that makes a good weather forecast.

b) Present your weather forecast to the rest of the class. Record your weather forecast on video.

c) Judge the presentation using the agreed criteria.

The Big Task

Is the weather the same everywhere around our school?

Have you ever wondered why some rooms in your new school are much warmer than others or why some places in the school grounds feel a lot colder and windier?

Take a number of weather measurements around your school for a week and record them in a table like the one below.

1. Draw a simple plan of your own school and its grounds like the one on page 76. Remember to include a title, key and north sign.
2. Mark on your map the locations where you took your weather readings (with numbers).
3. Describe what the weather readings tell you. For example, are some places warmer, colder, wetter, or more sheltered?
4. Explain why weather conditions vary around the school grounds.

		Temperature					Rainfall					Wind strength				
Site	Day	1	2	3	4	5	1	2	3	4	5	1	2	3	4	5
1. North-facing side of building																
2. South-facing side of building																
3. In the middle of games pitch																
4. In the middle of car park																
5. Close to trees or bushes																

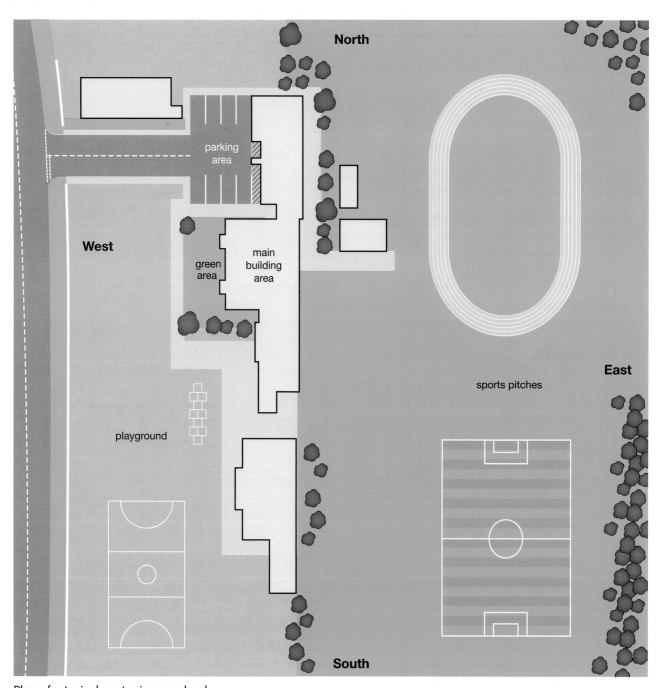

North

West

East

parking
area

main
building
area

green
area

sports pitches

playground

South

Plan of a typical post primary school

Fieldwork: A geographical investigation

Learning intentions

I am learning:

- the purpose of undertaking a geographical investigation
- to plan and carry out a fieldwork investigation
- how to use weather-measuring instruments
- how to present data using graphs and ICT
- to write up a report on the investigation
- to complete a self-evaluation of the geographical investigation.

What is fieldwork?

Fieldwork brings Geography to life. It will allow you to explore the world around you and this will help you gain a better understanding of geographical issues that you study in class. Fieldwork in Geography develops useful investigative, creative, thinking and writing skills. These skills will be useful in other areas of your schoolwork such as in Science and History.

Investigating the weather around your school

As you have seen, weather is an aspect of the natural environment that has a major impact on our lives. The weather patterns that create the climate of Northern Ireland help to shape the character of the place we live in. An investigation of weather patterns in the local environment will help you develop a better sense of the place where you live.

The aims of the investigation

For this field study, you are going to investigate the weather in your school grounds for one week. You will collect data to find out the relationships between different elements of the weather around your school. You will consider the following:

1. What happens when air pressure rises?
2. What happens when air pressure falls?
3. What happens when temperature rises?
4. What happens when temperature falls?
5. Which wind direction brings rain the most often?
6. Which wind direction brings the warmest weather?
7. Which wind direction brings the coldest weather?

Planning the project

1. Learn how to use and read the different pieces of equipment needed to measure the different elements of the weather.
2. If you can, use a digital camera to take photographs that can be used to illustrate the finished written report.
3. With your teacher, draw up a rota to involve everyone in your class in observing and measuring the local weather.
4. Decide on a convenient time to observe and measure the weather each day. Agree on how data for the weekend might be collected.
5. Record your observations and measurements using a **data collection sheet** (like the one below). You could change it to suit your needs.
6. When you have collected the weather data for one week you will need to write up a report outlining what you did and what you found out. Use the guide on pages 79 and 80 to help you.

Barometer

DATA COLLECTION SHEET

Investigating the weather around our school

Name: _____

Class: _____

Date of beginning and
end of investigation: _____

Time weather was recorded daily: _____

Day	1	2	3	4	5	6	7
Date							
Temperature (°C)							
Rainfall (mm)							
Wind direction							
Wind speed							
Cloud cover (1/8s)							
Air pressure (mb)							

Wind vane

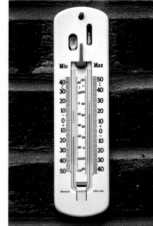
Thermometer

Rain gauge

Anemometer

How to ... complete the written report

1. Decide on an appropriate title for your investigation that describes what you were attempting to find out. This is called the **aim**.
2. Present your report under the following headings:

A. Introduction

Your introduction has two parts:

1 | The place where the investigation was carried out

Provide information about the place that you have investigated:

- Include a map of Northern Ireland to show where your school is located in relation to other important places. You could trace this from a map in an atlas or add the relevant information to a blank map provided by your teacher.

- Include a plan or sketch map (not to scale) showing your school and its grounds (like the one on page 75).

- Include photographs to help show what your area of study was like. Photographs are useful in providing a record of what you have observed or how you carried out a particular fieldwork task. They should be labelled and be referred to in your report.

- Draw a labelled **field sketch** showing the locations where measurements were taken and observations were made.

- Give your maps a title and include a key, a north sign and scale. Remember to number all maps, plans or photos that you supply, e.g. Map 1, Plan 1, Photograph 1.

2 | Geographical background

Give some information about the geographical background to your investigation, e.g. what you learned about the topic in class before undertaking the investigation and what you expect to find out. Is it true, for example, that low pressure brings a period of unsettled, wet and windy weather?

B. Method used for data collection

Outline briefly *how*, *where*, *when* and *why* you collected your data:

1. Start by explaining *how* you took the various measurements, including *what* pieces of equipment you used.

2. State the locations of *where* you took your measurements. Highlight them on the plan or sketch map of your school and its grounds.

3. Say *why* these locations were chosen as measuring and observing points. What do you already know about how weather can vary around a school building and grounds?

4. Note the time span that the investigation covered and state the time at which the data was collected. Explain why the time of year was chosen for the investigation and why the time of the day was chosen to collect the data in the field.

5. Include a completed data collection sheet and any relevant photographs to illustrate how the investigation was completed.

C. PRESENTATION OF RESULTS

Present your results using:

Tables: When data is set out in tables (including those produced from spreadsheets using ICT) it is easier to see and understand.

Graphs: These allow you to identify any trends or unusual values that do not fit the general trend. Try to use a selection of graphing techniques.

Other diagrams: Your findings about wind direction could be presented by a wind rose.

Hint: All tables and graphs should be numbered (e.g. figure 1/table 1), have titles and labels and keys where appropriate.

D. Analysis of the results

Analysis is where you describe what each of your tables of data, graphs and diagrams show. Begin by making simple statements about what each table, graph and diagram shows. It is important to quote actual numbers or values from the tables, graphs and diagrams. Refer to these by number in the written report.

E. Interpretation

Interpreting involves you trying to give reasons for your results. Try to identify any similarities, differences, trends and patterns in the data that you have collected, e.g. between different elements of the weather, such as rainfall and pressure.

F. ➡ CONCLUSION

The conclusion links the results to the aims of the investigation. Give answers to the questions posed at the beginning of the investigation. Refer to maps, tables and graphs in the report that support the answer. Relate the results to the geographical background in the introduction to the report.

G. Evaluation

This is your overview of the whole investigation.

1 Outline what you found to be the most interesting and enjoyable aspects of the investigation and what elements of the investigation went well and why.

2 Describe any weaknesses or limitations of the investigation. This might include difficulties you had in collecting the data, the nature of the equipment you used and the amount of data collected. Indicate how you tried to overcome any such problems.

3 What have you learned from completing this geographical investigation into the weather around your school? Are there any important lessons that you have learned that will help you when you have to complete a similar exercise in the future?

6 Farming today

A combine harvester on a modern farm in Northern Ireland

Learning intentions

In this chapter, I am learning:

- the value of farming to the Northern Ireland economy
- how farming has changed in Northern Ireland
- how to read and interpret trends using graphs and tables
- how to read and interpret an Ordnance Survey map
- how to make some decisions about the future of an individual farm.

What are farms like in Northern Ireland today?

Farming is an important industry in Northern Ireland. It has always been a significant contributor to the economy of Northern Ireland. It employs people and provides some of the food we eat. As you may already have learned, farming is a **primary industry** because it involves people working in jobs where they take things from the land.

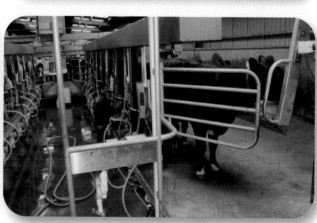

Get Active 6.1

Look at the photographs opposite that show what you might see on farms in Northern Ireland.
a) Describe what each photograph shows.
b) Explain how these photographs fit in with your image of what a farm in Northern Ireland looks like.
c) In groups or with the class, discuss what the photographs tell us about farming in Northern Ireland today.

Types of farms

Not all farms are the same. There are three types of farming:

- **Arable farms** concentrate on the growing of crops. There are few farms of this type in Northern Ireland.
- **Pastoral farms** concentrate on the rearing of animals, such as sheep on hill farms or beef and dairy cattle in more low-lying areas.
- **Mixed farms** grow crops and rear animals on the same farm.

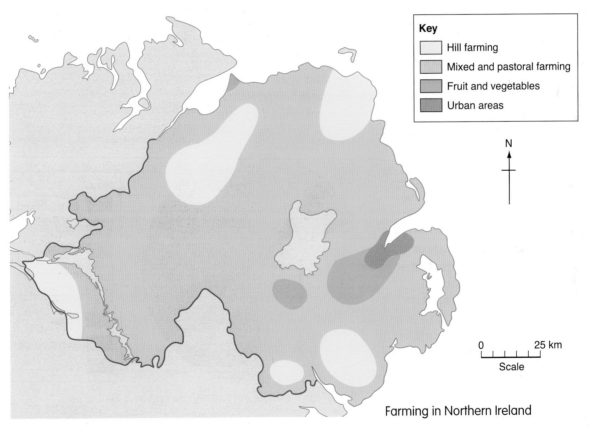

Key
- Hill farming
- Mixed and pastoral farming
- Fruit and vegetables
- Urban areas

N

0   25 km
Scale

Farming in Northern Ireland

Get Active 6.2

This is a simplified map showing the main types of farming in Northern Ireland. Use the map to describe the distribution of farm types in Northern Ireland. You may find it helpful to look at a physical map of Northern Ireland when completing this task. Try to include most of the following terms in your answer:

hilly	drier	east	arable farming
wetter	pastoral farming		hill farming
mixed farming		west	fruit and
vegetable farming		climate	flat land

How does farming affect our economy?

In 2004, farms in Northern Ireland provided 54,000 full- and part-time jobs. This accounted for 4.5 per cent of the total working population. (In the United Kingdom, the Republic of Ireland and the **European Union** (EU) the equivalent figures were 1.4 per cent, 6.0 per cent and 4.0 per cent.) By June 2005, the number of jobs in farms in Northern Ireland had fallen to 51,100.

Most farms are family owned. The average farm size is 36 **hectares**, and there are just over 13,000 farms that are large enough to provide full-time work for one or more persons. Many farmers in Northern Ireland today earn below the national minimum wage (£5.35 per hour in 2006 for workers aged 22 and above). Only 4 per cent of the farmers are under the age of 30, and the average age is 58. Farmers in Northern Ireland care for approximately 80 per cent of the land area.

In 2004, there were just under 28,000 farms in Northern Ireland farming slightly more than 1 million hectares of land. Farming was worth £1.23 billion to the Northern Ireland economy. The pie chart below shows how much individual farm sectors contributed to this total of £1.23 billion.

Get Active 6.3

a) Using the pie chart, rank the value of the different farm sectors in Northern Ireland in 2004 (from highest to lowest).
b) Using the information in the text above, calculate the percentage decrease in the number of full- and part-time jobs on farms in Northern Ireland between 2004 and 2005.
c) Compare the percentage of the Northern Ireland workforce employed in farming to that in the United Kingdom, the Republic of Ireland, and the EU.

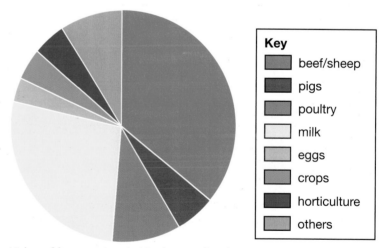

Key
- beef/sheep
- pigs
- poultry
- milk
- eggs
- crops
- horticulture
- others

Value of farm sectors in Northern Ireland in 2004

How has farming changed in Northern Ireland?

The kind of farming that existed in your grandparents' childhood no longer exists. When they were young, many people in Northern Ireland (including school leavers) depended on farming for their livelihood. Our countryside was a major source of employment and income, and went a long way to meeting our food needs. Consider the information in the following tables and graph, and then complete Get Active 6.4.

Get Active 6.4

Your task is to investigate some of the changes that have taken place in farming in Northern Ireland.

a) Using Table A, draw a line graph to show how the number of farms in Northern Ireland changed between 1952 and 2004.

b) Describe the trend shown by your line graph (quote actual figures).

c) Look at Table B. What has happened to the number of dairy cows in Northern Ireland between 1981 and 2004?

d) How has the **average annual** milk **yield** per cow (in litres) changed in that time?

e) Use the information in Table B to calculate the total amount of milk produced (in litres) in Northern Ireland in 1985 and 2004. (For example, in 1981 the total was 270,000 x 4,430 = 11,961 million litres.)

f) Look at the data in Table C. Describe what happened to the numbers of beef cows, sheep, pigs and poultry between 1981 and 2004.

g) Look at Graph D. Describe how the production of potatoes, barley, wheat and oats has changed in Northern Ireland between 1981 and 2005.

Year	Number of farms
1952	82,710
1961	71,200
1972	62,146
1981	40,724
1990	35,358
2004	27,614

Table A Number of farms in Northern Ireland
Source: DARD Statistical Data 1981–2004 and Ministry of Agriculture (NI)

Year	Number of dairy cows	Average annual milk yield per cow (litres)
1981	270,000	4,430
1985	297,000	4,590
1990	277,000	4,880
1995	272,000	5,240
2000	286,000	5,840
2004	290,000	6,270

Table B Northern Ireland dairy herd and milk output
Source: DARD Statistical Data 1981–2004

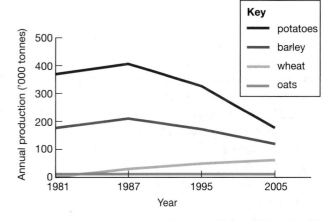

Graph D Crop production in Northern Ireland (1981–2005)
Source: Graph based on DARD Statistical Data 1981–2005

Year	Beef cattle	Sheep	Pigs	Poultry
1981	231,000	1,284,000	729,000	12,209,000
1985	227,000	1,787,000	719,000	10,720,000
1990	269,000	2,825,000	685,000	11,196,000
1995	309,000	2,754,000	637,000	16,003,000
2000	318,000	2,741,000	413,000	15,376,000
2004	296,000	2,225,000	424,000	20,509,000

Table C Number of beef cattle, sheep, pigs and poultry
Source: DARD Statistical Data 1981–2004

Farm size and farming methods

As you will have seen, the changes that have taken place in the farming industry in Northern Ireland have transformed the nature of farming. Many farmers have either sold their farms or have been forced to scale down their operation. There are two main trends in the farming industry in Northern Ireland, as follows:

Change in farm size

As the number of farms has declined, individual farm holdings have got larger.

Change in farming methods

In your parents' lifetime, farmers introduced more **intensive farming** practices. These involve greater use of chemicals (such as **fertilisers** and **insecticides**) on the land as well as the use of machinery. The main effect of intensive farming methods has been to change the appearance of the countryside. Such methods have led to:

1. Higher yields because more crops can be grown or more animals raised in the same area.

2. Fewer farm workers and more machines.

3. Larger fields due to the removal of hedgerows.

4. Cutting down of woodland to make more land available to farm.

5. Draining of wetland areas to make more land available to farm.

Year	Number of small farms	Number of large farms
1997	11,366	551
1998	11,362	608
1999	10,766	616
2000	10,547	656
2001	9,833	729
2002	9,388	759
2003	9,365	764
2004	3,966	1,310

Numbers of small and large farms in Northern Ireland, 1997–2004
Source: DARD Statistical Data 1981–2004

Get Active 6.5

Look at the table above.
a) Copy this data into a spreadsheet (such as Excel®).
b) Use the **chart wizard** to produce a line graph. Use two different colours to represent small and large farms (on the same line graph). Give your graph a title, a key and label the axes appropriately.
c) Describe the trends shown by your graphs.
d) Do you think the trends shown in your graph will continue in to the future? Why do you think these trends are happening? Give reasons to support your views.
e) In groups, discuss the advantages and disadvantages of farmers employing intensive farming methods.
f) On a large sheet of paper, record the views of the group in a table like the one below.

Intensive farming practices	
Advantages	**Disadvantages**

g) Report what your group agreed to the rest of the class.
h) In your own words, outline what you see as the advantages and disadvantages of farmers employing intensive farming practices.

Crisis in the farming industry

In recent years, our newspapers and television screens have been filled with stories about problems in the farming industry. Many food scares (such as BSE) can be traced back to the farming methods used in the industry. Farmers are often under pressure to supply high yields to remain in business.

BSE HITS NORTHERN IRELAND

'Mad cows' are for real!

Get Active 6.6

a) Your task is to investigate the stories behind the headlines on the right. You can use a search engine to look for related articles on the internet. Try www.google.co.uk or www.bbc.co.uk/news or the 'News' section on Learning NI. Key terms for your search could be: 'mad cows', 'BSE', 'avian flu', 'cheap food imports'. You could narrow your search by adding the term 'Northern Ireland' to the above terms.

b) Investigate each headline by answering the 'five Ws': What? Where? When? Why? Who? Make notes to answer each of the key questions.

c) Work in small groups. Compare your notes and agree on the key points that explain the original headlines.

d) Report back to the class what your group found out.

WILL AVIAN FLU REACH NORTHERN IRELAND?

Strong pound attracts cheap food imports

Organic farming techniques

One of the consequences of food scares is that people are now looking more closely at how their food is produced. Those who choose not to buy food that has been grown using fertilisers, **pesticides** and other chemicals, fear some of these harmful substances could enter the food chain or water supply and affect our health. People are becoming more conscious of the need for healthy eating, and this is reflected in the choices that they make when they buy their food.

Many are also concerned about the impact of intensive farming methods on the environment. These trends have lead to the growth of **organic farming** where the farmer does not use chemicals. Instead the farmer uses animal manure and **crop rotation** to keep the soil fertile. This type of farming is less intensive. It causes less harm to the environment because fields are kept small, preserving hedgerows, ditches and walls. However, the problem with less intensive farming methods is that yields are lower.

Cattle on Greenmount Organic farm in County Antrim

Get Active 6.7

This task will involve some independent research.

a) Go to your local supermarket or other food outlet, or look at a website of a supermarket.
Look at the range of organic foods on offer.
- List the different organic products on offer.
- Note where these products originated.
- Describe how organic products compare in terms of size and appearance to their 'normal' counterparts.

b) You have to buy the following items: 4 kilos of potatoes, 2 kilos of carrots, 1 kilo of tomatoes, 1 kilo of bananas, 3 litres of milk and 500 grams of cheddar cheese. Use a table like the one opposite to record the price of the organic products and their 'normal' counterparts.

c) Calculate how much you would have to pay if your basket contained only organic products, and then if your basket contained only normal products.

d) How much more expensive is the organic food basket?

e) What effect might such a difference in price have on the shopper?

Extension

You have been employed as an assistant promotional manager for 'The Northern Ireland Good Organic Food Company'. This company specialises in supplying organic food to shops throughout Ireland and the United Kingdom. You have been asked to design a poster encouraging people to buy organic products. You should outline the benefits of eating organic food and highlight the contrasts with food produced through extensive methods.

Food item	Normal price	Organic price	Price difference
4 kilos of potatoes			
2 kilos of carrots			
1 kilo of tomatoes			
1 kilo of bananas			
3 litres of milk			
500 grams of cheddar cheese			
TOTAL			

Farming and the European Union

When the **European Economic Community** (later to become the European Union) was established, one of its main aims was to increase food production in Europe. This would be achieved through the **Common Agricultural Policy (CAP)** which:

- put **tariffs** (taxes) on farm produce imported into EEC countries
- guaranteed the price for farm products (such as butter, milk, meat and wine) if more were produced than could be sold, and stored this **surplus** production (so that it could be used in years when less was produced)
- paid **subsidies** to farmers to bring more land in to cultivation (through draining wetlands, improving the quality of the soil and removing hedgerows to make fields bigger).

An outcome of the CAP was that farmers produced more of some foodstuffs than was needed. This meant vast quantities of food were kept in storage, and led to the creation of the **butter**, **beef** or **wheat mountains** and the **wine lake**.

An EU butter mountain

Get Active 6.8

a) In pairs, think about the good and bad effects of the Common Agricultural Policy.
b) Share your views with the rest of the class.
c) Record what you have learned in two lists.
d) Do you think farmers in Northern Ireland would have liked/disliked this Common Agricultural Policy? Explain your reasons.

An EU wheat mountain

Revising the CAP

In an attempt to reduce the **over production** of food in Europe, the CAP was altered to introduce:

- a **quota** system which established how much of a product (such as milk) farmers could produce
- a **set aside** scheme which paid farmers not to use their land to produce food
- **diversification** grants which provided money to farmers to use their land for other purposes (than the production of foodstuffs)
- support for **conservation schemes** which protected environmentally sensitive areas.

Get Active 6.9

a) With the class, discuss the following:
   'In a world where millions of people are starving is it a good idea to promote a set aside scheme?'
b) Design a logo for the front of a T-shirt that reflects your views on this topic. It might be a slogan, a cartoon or whatever. It should be colourful and attention grabbing.

Diversification: How farms can create alternative sources of income

Some farmers are finding it difficult to make a profit from their traditional farming activities. They are looking to find new ways of using their farm to make money. This is called diversification. This is where farmers seek alternative income, other than from growing food crops or conventional livestock rearing. The diagram below shows five routes in which a farmer might choose to diversify.

Source: Ulster Farmers Union

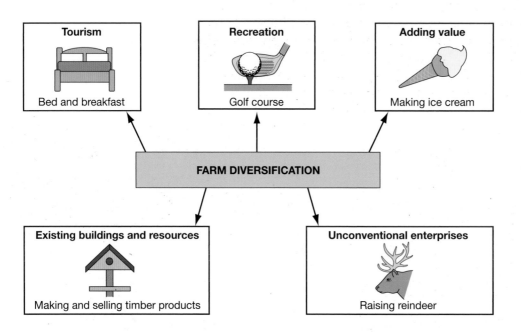

Tourism	Recreation	Adding value
Bed and breakfast	Golf course	Making ice cream

FARM DIVERSIFICATION

Existing buildings and resources	Unconventional enterprises
Making and selling timber products	Raising reindeer

Get Active 6.10

Think of the possible ways in which farmers might choose to diversify. Record your answers in a table like the one below.

Ways for farms to diversify				
Tourism	Recreation	Adding value	Existing buildings and resources	Unconventional enterprises
Bed and breakfast	Golf course	Making ice cream	Making and selling timber products	Raising reindeer

The impact of diversification on dairy farming

Around 10 per cent of farmers in Northern Ireland have diversified from their traditional farming activity. Against a background of falling farm incomes and increasing pressures on farming, this figure will continue to rise. Farmers in other parts of the United Kingdom faced similar challenges as their counterparts in Northern Ireland, and some have come up with ways that enable them to remain in the farming business.

In Buckinghamshire, the demand for milk has traditionally come from London and local **commuter** towns. Dairy farmers had a reliable **market** for their product and they farmed the land in an efficient and productive way. However in the last 25 years, the main change to affect **dairy farming** has been milk quotas. These were introduced in 1984 and limited the amount of milk that individual dairy farms could produce. Milk production in the United Kingdom has been restricted by the quota system, to the extent that today the United Kingdom is no longer able to produce all the milk that is needed. Milk production in the United Kingdom has dropped from being about 110 per cent to 86 per cent self-sufficient in milk. The price that the farmer receives for milk has also fallen. This has made dairy farmers look to the benefits of **adding value** to their enterprises, by diversifying into new areas, as their income becomes too low to earn a decent living.

A dairy farm shop which sells own-made ice cream

Get Active 6.11

a) Use an atlas to find the location of Buckinghamshire.
b) List the reasons why this is a good area to develop dairy farming (include factors relating to both the **physical** and **human environments**).

Diversification on Beechdean Farm

Diversification schemes can change the nature of farms and the everyday work done by the farmer. Beechdean Farm provides an example of what one dairy farmer has done to increase farm income. The farmer has chosen to go down the route of adding value to what is already produced, which is milk. As you will have seen in Get Active 6.10, there are five options open to a farmer looking to diversify:

- tourism
- **recreation**
- adding value
- using existing buildings
- **unconventional** enterprises.

Get Active 6.12

Check out the website of Beechdean Farm (www.beechdean.co.uk). Find out about the location of Beechdean Farm by using www.multimap.com and typing in the postcode of the farm: HP14 4NL. Here you can access location maps at different scales (1:5,000; 1:10,000; 1:25,000; 1:50,000, etc). Use the icons on the tool bar to look at aerial photos showing the location. Learn to toggle between the OS map extracts and the aerial photos of the area.

a) Use this information to write a brief description of Beechdean Farm, including its location.

b) Which map is of most use? Why?

c) What is the benefit of such **technology**?

d) Assess the impact of diversification on the farm and the farmer, by answering the following questions:

- What kind of image does the **home page** of Beechdean Dairies try to portray?
- Where is Beechdean Farm located?
- What **breed** of cow does the milk come from to make the ice cream?
- What are the main products of Beechdean Dairies?
- What kind of market is the Beechdean Dairies range of products aimed at?
- When did Beechdean Farm start to make ice cream?
- What do you think are the main changes on the farm?
- How do you think the job of the farmer might have changed?

The Big Task

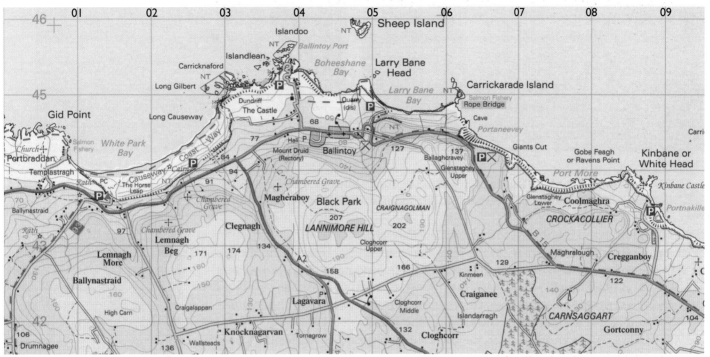

The McHenry family farms 118 hectares of land on the north coast of Northern Ireland, on a part-time basis. Due to the area's location and climate, sheep farming has been the predominant source of income on the farm. The farm can be found on the map extract for the area above. It is located at grid reference 0342. The farm surrounds the farmhouse and includes land on Black Park (0443).

Making a profit from sheep farming in Northern Ireland is becoming more difficult. Mr McHenry has to compete with cheap imports at a time when demand for meat is falling, due to the various recent health scares linked to meat.

Your task is to decide how Mr McHenry should best use his land to make the most money. As we have discussed, farmers are looking at non-farming activities to add to or even totally replace their traditional farm income. Mr McHenry also wants his farming to be *sustainable*.

Use the map extract to find out the following (use the key on page 137 if necessary):

1. Which bay does the farm overlook?
2. Which sandy bay is located to the west of the farm?

Sustainable farming means meeting people's need for food and other resources without doing harm to the environment. If farming is to be sustainable, it will have to:

- produce safe and healthy foodstuffs (as well as other products and services) to meet the demands of the public (such as: milk, meat, timber, energy crops, farmers' markets, farm shops)
- allow farmers to earn a fair and worthwhile income
- provide environmental improvements (such as restoration of habitats)
- provide other benefits that the public want (such as: **access** to the countryside, leisure and tourism amenities, nature reserves, accommodation)
- achieve the highest standards of animal health and welfare.

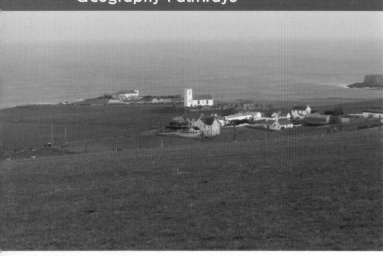

3. Which road would the farmer take to go to Ballintoy?
4. How far, and in what direction from the farm, is the church in Ballintoy (each square = 1 km)?
5. What is the height of Lannimore Hill?
6. What evidence is there on the map that tourism is important in the area?
7. What evidence is there on the map that people have lived in the area for a long time?
8. What do the contour lines between Lannimore Hill and the coast tell you about the nature of the land on the farm?
9. Using map evidence, describe the contrasting coastline along this part of North Antrim.
10. Describe the land found on the farm using the photographs above and on the right.
11. Work in groups. Discuss how Mr McHenry might diversify. Decide on the best way forward for his farm in the future.

   • Use information from the map.
   • Think about what you already know about this area of Northern Ireland to support your views (such as proximity to Portrush, Portstewart, Coleraine and the Giant's Causeway).
   • Look back to what you learned about the different diversification routes that a farmer could follow.

12. Make notes on how Mr McHenry could diversify:

   • Outline the changes he would need to make to his farm.
   • Identify the customers for the new products.
   • Describe how the proposed scheme will benefit the farmer and the environment.

13. One member of each group explains what they have decided to the rest of the class.
14. List all the proposed diversification schemes drawn up by the whole class. Rank the schemes in order of how likely you think they are to make this farm more sustainable in the future.
15. What do you consider to be the most important things that you have learned about how farming in Northern Ireland is changing? You could show this in a concept map.

7 How much do we know about our food?

A family dinner

Learning intentions

In this chapter, I am learning:

- about where our food comes from
- about our changing food tastes
- about the consequences of choosing to eat the food we do
- about the significance of 'food miles'
- how to calculate distances on maps at a range of scales
- to implement a Fairtrade policy in my school.

What do you eat?

Shopping for food, cooking food and eating food are important aspects of our lives. However, what we buy and what we eat affects not only ourselves, but a number of other people, such as shop keepers, **wholesalers**, packagers, transporters, processors and farmers. The food choices that we make raise important questions for other people: the 'rights and wrongs' about the way we produce food, and the effects of consuming certain types of food on the environment. This is why the decisions that we make about the food we eat are felt at **local**, **national** and **global** scales.

Food glorious food

On an average day, most of us eat a variety of food items. Have you ever stopped to consider where all this food comes from? Our food links us, and our community, to places all over the world!

On the next page is an example of what John, a Year 8 student, eats almost every Friday.

FRIDAY

Breakfast: cereal (cornflakes) with milk and sugar, 2 slices of toast, mug of tea

Break-time: bag of crisps

Lunch: pizza, chips, can of cola

4.00 pm (arrival from school): 2 cheese sandwiches, mug of coffee

Tea-time: chicken curry, can of cola

10.30 pm (after cinema): burger meal (large burger, chips, cola)

Get Active 7.1

Through our food we are connected to many different countries in the world.

a) Make a list of ten items in your fridge/freezer/or cupboard. Find out where they came from. (Hint: Look carefully at the labels or packaging.)

b) Work out the average distance from their countries of origin to your fridge. Use an atlas to help you do this. Record your answers in a simple table.

Extension

a) Collect some food packaging or labels.

b) Design and plan an effective visual presentation to highlight where our food comes from. (You might choose to use some of your food packaging or labels in your presentation.)

THINK!

- Think about how to do it.
- Jot down your own ideas.
- Discuss your ideas with a partner.
- Share these with the class.
- Agree on the best method.

Where does our food come from?

Today most of the food we eat is bought in supermarkets. Some people even choose to buy their food over the internet from home. In either case, a vast range of food items can be found. However, most people today know very little about where the food they eat comes from.

Are all these fruit and vegetables grown in the United Kingdom?

Get Active 7.2

a) Look at the food packaging and labels from the previous exercise. List all the countries where pupils' food came from.

b) Sort these countries by **continent**.

c) Classify the countries under **more economically developed country (MEDC)**, such as the United Kingdom or France, or **less economically developed country (LEDC)**, such as Mali or Pakistan.

d) What conclusions can you draw about where much of our food comes from?

Food miles

Food miles are a measure of the distance food travels: from where it is grown to where it is bought (or eaten) by the end-user.

Before engines were invented, most food came from the local area near to people's homes – this was no further than someone could walk, or a horse could pull a cart to market. This meant that little or no **fuel** was used to transport food. Today, as you saw in the last exercise, the food you buy comes from all over the world. Much of your food probably travels thousands of miles before you get it. Food miles tell us how environmentally friendly a food item is. Generally speaking, the lower the food miles, the better choice the product is for the environment because it has generated less pollution. Basmati rice, on sale in one of Northern Ireland's supermarkets, is grown in India, packaged in Italy and stored in England before being sent to outlets in Northern Ireland. This is true of many food items that we eat.

Melons from Spain   Oranges from Israel

Apples from France   Potatoes from Cyprus   Tomatoes from Italy

The origins of different fruit and vegetables in a supermarket in Northern Ireland, and (right) some of the different forms of transport they would have taken to get here

What is your favourite meal?

It can be difficult to find locally grown food today. For example, unless you have grown your own vegetables or fruit, you can't be sure how far your food has actually travelled to get to your local supermarket. We are going to look at where the food eaten by all the pupils in your class for their favourite meal has come from.

FISH POTATOES PEPPERONI CHICKEN OLIVES RICE TOMATOES CHEESE BEEF BREAD LETTUCE

Get Active 7.3

Everyone has a favourite meal. It might be fish and chips, pizza, beef curry, chicken tikka, traditional Sunday dinner or something else.

a) Name your favourite meal. List the individual food items or ingredients.

b) Identify the country of origin of these items/ingredients.

c) On a blank world map with a scale:
- locate and mark your home
- locate and mark the sources of the food items/ingredients
- give your map an appropriate title and key.

d) Calculate the distance in kilometres from the sources to your plate at home:
- use a ruler to measure the straight-line distance between the source of an item and your home on the map (in centimetres).
- using the scale bar on the map, convert this to kilometres.
- complete a chart (like that for John below) to show the food miles for your favourite meal.

Item	Source	Food miles (kms)
Rice	USA	5,900
Beef	Argentina	11,400
Curry powder	India	7,000
Cola	Lisburn, Northern Ireland	50

John's favourite meal

e) List all the countries identified by all the pupils in the class.

f) If you limited your favourite meal to food originating from Ireland (north or south), would the food miles have differed significantly?

Get Active 7.4

Create a 'local' dish. This means choosing only food items or ingredients that originate in Ireland:

a) List all the food items/ingredients that make up this dish.

b) Use a map of Ireland (with a scale) to:
- locate and mark your home
- locate and mark the sources of the food items/ingredients
- give your map an appropriate title and key
- calculate the distance in kilometres from the source to your plate at home.

c) How different is this 'local' meal to your 'favourite' meal (in terms of food items and food miles)?

d) Would your 'local' meal be affected by the time of the year that you were cooking it? How might you solve any problems relating to this?

Changing tastes

The eating habits of people in Northern Ireland have changed greatly in the last ten years. This is reflected in the products that we find on the shelves of our supermarkets.

A 1950s grocer's

A supermarket in the present day

Get Active 7.5

Work in groups of three.

a) Talk about what the two photographs below tell us about what a typical family in Northern Ireland might have eaten 50 years ago. Consider:
- What might the children be eating and drinking?
- How does the food differ from what we eat today?
- Would such food have been more/less healthy? Why?
- How does food in the cupboard differ from what might be found at your home today?

b) Based on your group discussion, write a description of a typical meal eaten by a family in Northern Ireland in 1950.

A family meal in the 1950s

The contents of a 1950s kitchen cupboard

Factors affecting our food choices

Many factors have combined in recent years to help us change our food choices.

Cheap air freight

Year-round availability of fruit and vegetables, even when they are out of season in Northern Ireland

Increasing number of Cookery books in our shops

Improved refrigeration methods

Improved technology allows produce to be ordered and delivered within 24 hours

Cookery programmes on television

Arrival of immigrants with different tastes in food

Arrival of major supermarket chains in Northern Ireland

Improvements in farming methods in LEDCs

Get Active 7.6

You will need a set of cards and a copy of a diamond frame.
a) In pairs, think about the factors that have helped change our food choices.
b) **Diamond rank** the cards (see opposite) with your partner.
c) Individually, show this information in another way. You could write an account of what you have learned, list the factors in a rank order, draw a concept map, or produce an information chart.

How to ... diamond rank

x ——— most important

x   x——— next most important

x   x   x— average importance

x   x——— next least important

x ——— least important

The life and times of a banana

There are many stages between the growing of a food product and its final destination. The journey of a banana starts with the **primary producer** in the **Tropics**, goes through **shipping agents**, wholesalers and **retailers**, before it gets to our lunchboxes or dinner tables in our homes and then the final resting place of the banana skin! The journey is very long, and involves many different people as can be seen in the photographs below.

The life of a banana

Get Active 7.7

a) With a partner, discuss and agree what each of the photographs show.

b) From what you can see in the photographs, draw a flow chart that records the life of a banana. (Remember to add appropriate labels to your chart.)

How does shopping in a supermarket affect food miles?

The food miles for products grown locally in Northern Ireland, such as fruit or vegetables, can differ enormously depending on whether they are sold by a local greengrocer or in one of the major supermarkets.

The Bramley apples from this orchard in Loughgall in County Armagh are sold in two different shops in the nearby town of Dungannon.

The apples sold in the greengrocer's shop are transported directly from the orchard near Loughgall to the shop in Dungannon. The apples sold in the supermarket are transported from the orchard to a packing house and then a distribution centre in Belfast, and finally to the supermarket in Dungannon as well as many other supermarkets.

Get Active 7.8

a) Look at the map of Northern Ireland on page 109.
b) Calculate the food miles for the County Armagh apples if they were to be purchased in the greengrocer's shop, by a shopper living in Dungannon.
c) Calculate the food miles for the County Armagh apples if they were to be purchased in the supermarket by a shopper living in Dungannon.

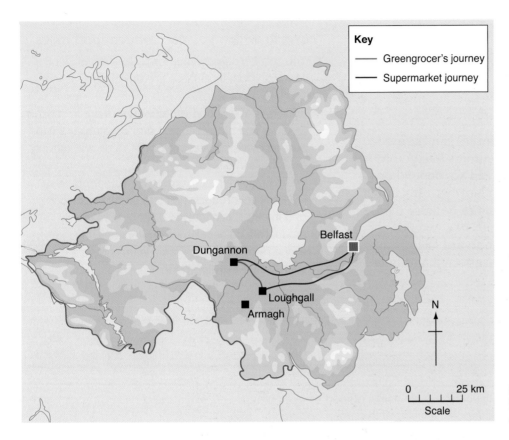

Key
— Greengrocer's journey
— Supermarket journey

Belfast
Dungannon
Loughgall
Armagh

N

0   25 km
Scale

Map of the different journeys for the County Armagh apples

Get Active 7.9

a) In pairs, discuss what you think are the **consequences** of eating food that has a high food miles value. Think about what the advantages are of buying and eating locally produced food and reducing the food miles. Advantages include:

- less packaging
- job creation
- encourages people to remain in the countryside
- reduction in noise pollution
- reduced carbon dioxide **emissions**
- rural regeneration
- less damage to roads
- encouraging **entrepreneurship**
- less traffic congestion.

This is not a complete list. You might think of more!

b) Use a table like the one opposite to help organise your thoughts.

Environment	What are the positive impacts of buying local food on the local and global environment? • • •
Economy	How does buying local food affect the local economy? • • •
Society	How does buying local food help the lives of local people? • • •

Food miles ... an issue

Once we begin to consider **concepts** like food miles, a number of important questions begin to emerge.

Reducing food miles

The desire to reduce food miles can raise difficult questions for consumers like you in MEDCs, when farmers in LEDCs depend on our **consumption**. If we in Northern Ireland decided to stop eating bananas because they have a high food miles tariff, how would banana growers in the Windward Islands (an LEDC in the Caribbean) be affected and how might they feel?

Ways in which *you* can reduce food miles

- You can encourage your family to shop at local farmers' markets, if there are any close to where you live.
- You can eat fruit and vegetables that are in season locally – find out which fruit and vegetables are in season by visiting www.bigbarn.co.uk and clicking on 'What's In Season'.
- You can look for local produce in supermarkets. Buying local products helps to protect local jobs and the local economy. If buying more locally produced food benefits the environment, the economy and society, then it is an example of **sustainable development**.

Get Active 7.10

Work in groups of four.

a) Appoint a recorder who will note the main points agreed by each group.

b) Appoint a reporter who will report the views of the group back to the class.

c) If we in Northern Ireland decide to stop eating bananas because they have a high food miles tariff, how would banana growers in the Windward Islands be affected and how might they feel? You need to mention:

- the location of the Windward Islands (use an atlas)
- why the Windward Islands is termed an LEDC
- effects on the economy
- effects on the people
- effects on rural areas
- how the people on the islands, including children, would feel.

d) Record what you learned in this exercise on a drawing of an outline of an islander:

- Note the effects around the outside of the outline of the islander.
- Note the feelings inside the outline of the islander.

Fairtrade

Once you start looking at locally produced food available in your local supermarket, you will soon realise that it is not possible to grow all the food we like in Northern Ireland. However, we can still make a difference if we think carefully about what we buy. For example if we buy bananas with a '**Fairtrade**' sticker, this means that we are making sure that the farmers who produce the bananas are getting a fair deal from the big food distributors and supermarkets. It also means they are using environmentally friendly methods to grow their bananas.

Get Active 7.11

List at least six Fairtrade food items you could buy other than bananas. You could look in your local supermarket, use a search engine on the internet such as www.google.co.uk, or visit specific websites such as www.fairtrade.org.uk, www.oxfam.com and www.actionaid.org.uk.

Being a global citizen: Getting involved in Fairtrade

As you may have realised from a visit to your local supermarket, an increasing number of Fairtrade products can be purchased in Northern Ireland. You can now spread the Fairtrade message to your friends and family, highlighting the benefits it brings to the people of less economically developed countries. You can try to influence others in your school.

> Being a global citizen is about how you choose to live your life. Do you stand up for people living in poverty? Do you care about whether your actions make life better or worse for other people? Do you think about the products you buy and whether the people who produced them are getting a fair deal? Do you want to get involved, in even the smallest way, in making the world a better place?

Source: Oxfam/Cool Planet, 2006

Get Active 7.12

Your task is to spread the Fairtrade message to everyone in your school, highlighting the benefits it brings to the people of LEDCs. You need to:

a) In groups, discuss how you will produce a poster for an agreed class  **charter** about buying Fairtrade products:

- Find out what a charter is. Look at an example on the internet, how it is set out and the kind of language used.
- Agree on what should be included in your charter.
- Keep a record of all the points agreed.
- Discuss your ideas with the class.
- Agree on a charter that covers all the points.

b) Individually, produce your own posters.

c) Place your class posters on display areas around the school.

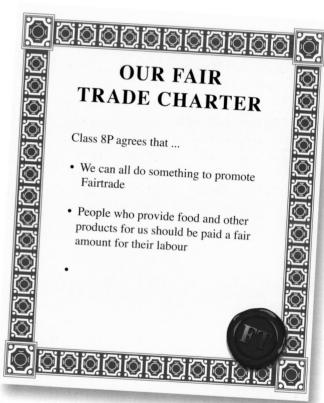

An example of a charter

The Big Task

Your challenge is to try to persuade the Principal and the school management to implement a Fairtrade policy in your school. This will involve you and the rest of your class in an Action Project.

Work in groups of four.

1. Identify areas where your school could use Fairtrade products.
2. Agree actions that your school could implement to promote Fairtrade.
3. Suggest events that the school could run to promote Fairtrade.
4. Feedback to the rest of class.

Then work as a class.

5. Agree on the best ideas, and then refine these ideas.
6. Decide on how to present your ideas. Then present your ideas to the principal.
7. Try to get an agreement on what steps the school will take and when. Implement the actions and run your event.
8. Talk about what has been achieved.
9. Write a short piece on what you have learned from the process and your participation in it.

Chew the facts!

Next time you sit down to a meal think about the following:

- Transporting food by air uses nearly 40 times the amount of fuel that sea transport uses.
- Food in the United Kingdom now travels 50 per cent further than it did two decades ago.
- A traditional Sunday lunch made from imported ingredients generates 650 times more carbon emissions than the same meal made from locally grown ingredients.
- The five largest supermarkets in the United Kingdom account for 80 per cent of the food market.
- In the 1970s, supermarkets offered about 5,000 different products compared to 40,000 today.

Source: *The Guardian*, 24 June 2004

What is GIS?

Learning intentions

I am learning:

- what GIS is
- to use GIS to find out about places and to create layers.

Geographical Information System (GIS)

Imagine a big pile of maps neatly stacked up. You can take a map out of the pile, or put another one in. You can see one map through another map if you want. You can combine as many maps as you want. You can put in your own maps. You can ask questions and get answers. You can do all of this with **GIS**.

GIS stands for Geographical Information System. It is information about places. It can be information about physical Geography (such as rivers, crops, roads, mountains or soil), or it can be information about human Geography (such as health, poverty, crime or buildings).

Buildings

Contour lines

Roads

Rivers

We use a GIS system to make sure we take the fastest route to an emergency.

I use GIS when I am deciding where to move my troops.

We use GIS to work out the best places to put our **mobile phone masts.**

GIS can be used to forecast what crops will grow best this year.

GIS is needed to protect rare species of plants and animals.

We can find leaking pipes using GIS much more easily than using paper maps.

Get Active 1

a) What do the statements above and opposite tell you about what GIS is?

b) Think about who would say these things about GIS, and explain how GIS is useful to each person.

c) Share your thoughts with a partner – do you agree with each other?

Using a GIS

One example of how a GIS is used is by the Environment Agency for England and Wales (www.environment-agency.gov.uk). It uses GIS to show, for example, the **water quality** of rivers, waste and **flood warnings**. Their website allows people to enter their postcode or town and view some geographical information.

To search for areas which could easily flood, click on 'Flood' in the 'Find Out About …' section of the website. Then put in a postcode in the 'Flood Map' section. You could enter 'KT1 1HR' (Kingston upon Thames, London) or put in a place name such as 'Hull' to see how it works.

Layering maps

Geographic Information Systems allow you to layer maps. This means that different information from different maps can be seen at the same time. For example, the Environment Agency site allows you to have a layer showing places and another layer showing **pollution hazards**. Why would they be useful? Can you think of other combinations of maps that would be useful in this way?

Get Active 2

a) With a partner, make a list of different things a map could show about:
- the quality of river water
- waste
- floods.

b) Go to the Environment Agency website with your partner (www.environment-agency.gov.uk) and make some maps using the GIS program. Remember to click on 'view map of these results'. Does the program give more or less information than you thought?

Get Active 3

a) Find a map of the area around where you live (using www.multimap.com) and print it out. Stick it in the middle of a large sheet of paper.

b) Work with a partner. Write examples of the type of people who might use GIS around the map, and include the advantages that computer-based maps would have to them over paper-based maps. Here is one to start you off:➡

> **Ambulance driver:** GIS can work out the fastest route from an accident to the hospital.

c) GIS is being used to try to reduce road accidents in parts of the United Kingdom. Accidents are put on a GIS layer and can be compared with other layers showing, for example, speed limit signs. This can help to identify the worst places for accidents. With a partner, discuss how GIS could be used to:
  * catch burglars
  * decide where a new school should be built
  * decide where money should be spent on a youth club
  * work out where an outbreak of a disease started.

d) Think of one other way for GIS to be used, and give this to another pair to work out. Make your idea sensible!

Extension

GIS can be used in many ways by different people. However, all GIS programs should be easy to use and have a number of useful features.

a) With a partner, set up a scorecard that would allow you to judge how good a GIS site is. You might want to judge it on the speed of the site, how colourful it is, how easy it is to use, how many features it has, and so on. Aim for about six features. You can score each of them out of 10.

b) Examine a range of GIS sites on the internet. You could try the ones below, or your teacher might give you some different ones:
  * www.multimap.com
  * www.mapquest.com
  * www.earthalbum.com

(Notice that they do different things, but you are judging how well they do them.)

A Geographical Information System: InfoMapper

One GIS system on the internet is InfoMapper. One really good way to use it is on an interactive whiteboard, if your school has one.

When you login to InfoMapper you will see details about the resources and projects you have and the groups you are part of. If this is the first time you have used InfoMapper there is likely to be little there.

When you first login you will see a map of Northern Ireland. It is called a 1:50,000 map. This is because a paper version of this map shows everything 50,000 times smaller than real life – it has been shrunk down 50,000 times. If one place is 1 centimetre away from another place on this map, in real life it is 50,000 centimetres away from it. How far is that in kilometres? This is the same scale as the OS map on page 136.

How to … navigate on InfoMapper

Here are some navigation techniques:

- **Moving around a map:** Drag the centre line on the screen to the left to see more of the map. Drag it back again to get back to the details on the left.

- **Zooming in and out:** You can zoom into the map by selecting the magnifying glass from the menu bar and clicking and dragging (hold down the left-hand mouse button and drag the mouse) the map towards the bottom of the map. To zoom out again, drag the other way. There are other ways of zooming in and out when you have selected the magnifying glass found just beside it. Work out with a friend what each of these does. You can use the hand on the menu bar to pull the map in any direction you want. If you get 'lost' because you have zoomed in or out too far, use the 'reset extents' button and the map will go back to what it was before – the fourth button on the bar above.

- **Finding a place:** You can jump to a place if you know its name or postcode. Find the 'Jump to' part of the screen, enter a place or postcode and then press 'GO'! This checks all the places that match the place you have put in. Pick the one that you want and you will be taken straight to it.

- **Showing more than one layer:** As you know, GIS systems use layers. You can use the 'Select a Map' option to pick another layer to show. You could choose the 'OSNI Aerial 25 cm layer' or some other OSNI map. If you chose the 'Aerial Photograph', this will open the Ordnance Survey of Northern Ireland (OSNI – the organisation in charge of drawing maps here) layer, with photographs of Northern Ireland taken from an aeroplane, which is why they are called aerial. They are called 25 cm photographs because anything bigger than 25 cm shows up as a pixel (or little coloured dot) on the photograph.

  When you have two layers at A and B, you have to tell the computer to put one on top of the other. Do this by clicking on the 'Layer the maps' button – the third button on the bar above. One layer is now on top of the other. Now you can drag the slider between the two layers to move between one and the other. Some of the maps might have white spaces if the digital maps for those places are not finished yet.

This all sounds difficult when you are reading it, but you will quickly get the hang of it when you actually do it. You could experiment with other maps and show your partner what you have found. There are maps that show, for example, the whole world on InfoMapper. Only OSNI maps will layer with other OSNI maps. The whole world layers will work with other world layers.

Other things with maps

You can do other things with maps that slide between two maps. When you have put one layer on top of the other, you will see a number of possibilities open to you to the left of the main menu bar.

This button allows you to place one layer beside the other, like in a mirror. Try zooming in and out and moving the map from side to side. Notice that the image is the same on each side.

The next button lets you stitch the layers so that, as a layer ends at one side, the other layer shows where the map continues. You can also move the maps on each side separately by using this button. If you want, you can have a small map on the top left-hand corner to show you where you are. Click on single map to get rid of it.

The left-hand button allows you to go back to the map, as it was at the beginning.

Viewing and creating layers

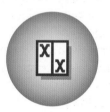

There are layers other than maps and aerial photographs that can be viewed with InfoMapper. Click on the 'Library' tab at the top of your screen. InfoMapper also allows you to create layers on the maps and to share them with others.

Click on the 'Add a Layer' button and follow the instructions. You will have to upload a digital file such as a photograph or a word processed document, but it could be a small video, a spreadsheet, a website, a diagram or any file that a computer can read. Then you can attach your file to the correct place on the map. Now try Get Active 4.

Jobs with GIS

If you are interested in Geography, especially maps and aerial photographs and computers, there are lots of interesting jobs you can do in GIS. It is likely that jobs in GIS will be even more important in the future, as computers get even faster and can handle even more information. Find out about jobs with GIS in Get Active 5.

Get Active 4

Work with a group to make layers on InfoMapper.

a) Your local area

You can gather lots of digital information about the place where you live and put it into a layer or a number of different layers in InfoMapper.

These could be about place names, or about local stories or about the industries you have in your area.

You could take digital photographs, make drawings on paper and scan them, record local people speaking place names or telling stories about places, put in graphs, websites or anything else which will tell people about your place. When you have all the files you need, upload them to your Resources in InfoMapper and link them to areas, lines or points on the map. Make sure you are as accurate as possible. If your layer is published, it will be seen by everyone who is able to use InfoMapper, including all schools in Northern Ireland and many schools all over the United Kingdom.

b) A journey

Everyone in their journey to school passes places that are important to them. These could be buildings or trees, streams or hedges. They could be places that are pleasant or places that are a bit scary. They could be busy road crossings or beautiful parks.

As a class you could identify these important places in your journeys to school, take photographs of them and write something to describe what they are, and explain why they are important to you. These files can then be attached to hotspots on the map.

c) Something on a bigger scale

You can create layers on maps which are not just in your area. You can even create layers on a world map. There could be links to all countries in which you and your classmates have relatives. You could have hotspots in each of them with details of their names and what they do. You might want to write something about the countries where some of the people who live in your area, or their families, originally came from.

You could check out the world layers on the library like 'Earthquakes' and 'Webcams' to see what others have done.

Get Active 5

a) Use a search engine to research GIS jobs. You could type 'GIS jobs UK' and see how many results you get, and compare this with other countries. Pick three of the job sites that are clearest and note the different types of jobs and, if the information is given, how much they earn and where the jobs are.

b) Make a page on your 'My Sites' area of LNI using the Advanced Wizard to give examples of GIS being used on the web. Use images where you can to make the page more exciting and link it to your index page so that others can see it. You could also put in links to some of the GIS websites you have used.

Geography in the news: What's it all about?

Learning intentions

I am learning:

- to ask the questions to find the Geography behind the news
- to question information given to me
- to produce a Geography news sheet.

How do we learn Geography?

You may think that you can only learn Geography at school, but there are many places to learn about Geography in the media: television, radio, internet, magazines and newspapers. You just need to look around you and ask some big questions (the 'five Ws'): What? Where? When? Why? Who?

Get Active 1

a) Read the article below about the Dalai Lama's visit to Northern Ireland on 20–22 November 2004 and answer all the Geography questions in the surrounding boxes. You are:
- questioning information given to you; and
- thinking how and why news events are occurring.

b) Try this with other pictures and articles that appear in the media using the same kind of questions: What? Where? When? Why? Who?

What's it all about?		
I. Who is the Dalai Lama?	2. Where does he come from? Where does he live now?	3. What happened to make him leave his home country?
4. Why did he visit Northern Ireland and what can we learn from the Dalai Lama?	### Dalai Lama's message of peace lingers The Dalai Lama spent three days in Northern Ireland this week, visiting reconciliation, community and religious groups. The 70-year-old exiled leader of Tibet, an international symbol of peace and non-violence, said that he wanted to "express my congratulations" that society here is more peaceful now than during his first visit in 2000 … One of his key messages was for compassion to be encouraged in Northern Ireland society … "Victims need special care … Compassion brings patience, for forgiveness … If you keep hurt for the rest of your life, that will destroy your life." 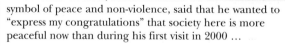 Adapted from *The Irish News*, page 16, 24 November 2005 5. When did he receive the Nobel Peace Prize and why did he receive it?	

Get Active 2

a) Read the article below carefully. It appeared in the *Belfast Telegraph* on 21 November 2005. Think about it as a Geographer!

b) Now record your 'What? Where? When? Why? Who?' questions and write down your answers to them.

Some answers will be found in the text of the article. For answers to other questions you will need to search in other places: on the internet, or on CD-ROMs that provide information about our world, or in an atlas or in the library.

Extension

a) Look through newspapers and find articles that relate to Geography.

b) Use a computer to create your own 'What's it all about?' recording sheets.

c) Read two of the articles that you find most interesting. Then record the geographical questions that you would like answered using the 'five Ws'.

What's it all about?		
Who?	**Where?**	**What?**
Why?		And then think of consequences in the short, medium and long term …

Marine food chain link being affected by global warming

Irish Sea plankton, a major link in the marine food chain, is being affected by global warming, according to a new WWF report.

It is also claimed certain fish, including John Dorys, usually found in waters off the southwest coast of the United Kingdom, are now being caught in Northern Irish waters.

Fish are increasingly threatened by the effects of **climate change** as temperatures rise in rivers, lakes and oceans, the report adds. It says hotter water means less food, fewer offspring and even less oxygen for marine and freshwater fish populations …

"We must act urgently on both emissions and fishing to protect fish populations as they are one of the world's most valuable biological, nutritional and economic assets."

Adapted from the *Belfast Telegraph*, 21 November 2005

When?

How to … report Geography news

Reporting on events involves producing a news sheet that provides interesting, easily understood Geography news from all over the world. This was a task that Rachel, a Year 8 student in a local school, was asked to do. She was asked:

- to collect articles from newspapers and from the internet
- to read them
- to put them into her own words
- then to present them in an illustrated *Geography in the News* sheet, using Microsoft® Publisher, or another desktop publishing software.

This was a big task. Her plan of action was as follows:

- Collect the articles over a period of time from local and national newspapers, and always record the date of publication.
- Classify/sort the articles into categories. Categories could be environmental, social, economic or maybe local, national, international. The categories could also be to do with natural events, or events caused by man.
- Select the most interesting articles.
- Decide on the articles that are the most easily understood.
- Write a summary of these articles.
- Find pictures/maps on the internet to illustrate the articles.
- Use Microsoft® Publisher to produce the news sheet.

Rachel reviewed her work. The tasks that she enjoyed were:
- searching for stories
- selecting pictures and maps
- designing the news sheet with illustrations, headlines, slogan, quotes and special points of interest.

The tasks that she did not enjoy so much were:
- putting the information into her own words (there was a temptation to copy directly from the article)
- arranging all the material – text boxes, maps, pictures, headlines – to make them 'fit'
- working out how to use Microsoft® Publisher (as she had never used it before).

Improvements that she would make, if she was doing this task again, were:
- to use Word Art for headlines
- to use different colours.

She thought that the following things would make the task easier:
- to be given guidelines on what is included in 'Geography'
- to be given a step-by-step guide to using Microsoft® Publisher (and maybe a class session to practise)
- to be given the criteria that should be used for the news sheet.

These are two pages from Rachel's first issue of *Geography in the News*.

GEOGRAPHY IN THE NEWS

NOVEMBER 2005   £0.75

GEOGRAPHICAL ISSUES FROM ALL OVER THE WORLD

VOLUME 1 ISSUE 1

INSIDE THIS ISSUE

Mount Everest is not as tall as we thought it was

The World's tallest mountain, Mount Everest, is shorter than we previously thought, a team of 50 Chinese experts confirmed yesterday.

Melting ice caps create a cash flow for Mr Broe

Global warming is considered to be terrible by environmentalists, but for Mr. Pat Broe it just means millions and millions of dollars. The melting ice caps are generating very good business for Arctic entrepreneurs,

but nobody has more of a reason to celebrate than Mr. Broe.

Mr. Broe is a railway magnate, but now he has a new job. He bought the isolated and frozen sub-Arctic Canadian port of Churchill, Manitoba,

in 1997, for $7 (£3.90). Now, because of global warming, the ice caps in Churchill have melted, creating

a new shipping season for the Arctic port. This means millions of dollars for Mr. Broe.

WEATHERWATCH

With satellite images and computer models, items such as weather vanes seem like scrap metal for measuring the weather. Although they used to be useful for telling the weather patterns, technology has taken over. Computers and satellite have proven to be better devices as they are more

accurate and are easier for forecasting the weather, the meteorologists claim. Yet weather vanes were essential for the largely farming societies which relied on them to predict weather patterns. Many still use the weather vane for their own needs, particularly people in isolated areas such

as farms in the countryside or prairies.

SPANISH DROUGHT

Spain had its first drizzle of rain in over a year, but not as much as it really needs. The rain fell in parts of south, western and central Spain. The average rainfall was only 1 mm but heavier falls are forecast for

this week. It has been the worst drought since 1947. Meteorologists say that the rain won't fall on the places that need it most, in Murcia, Alicante and eastern Andalucia. Meanwhile the

western parts of Spain were bracing themselves for Hurricane Vince.

SLOW RATE OF ASSISTANCE FROM GOVERNMENT, VILLAGERS COMPLAIN

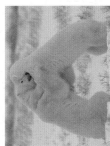

In Kashmir, the angry villagers complained of slow assistance from the Indian government. They were protesting against the slow pace of rescue and aid efforts reaching them. They blocked the main road for 45 minutes, shouting and pleading for help

against the earthquake that wrecked their lives.

The Indian army flew in plane loads of tents, food and water, but apparently it is not enough. People are saying that the aid workers and press are just driving by and taking no notice of them.

Getting involved in the world of news

Learning NI is a resource that you can access to really get involved in the world of news. Click on the news button on the LNI gateway (this part of LNI provides interesting news stories, loads of useful facts, opportunities to respond to surveys, tips for writing and more!). Have a look. Take a bit of time to explore what is on the site and see if there are particular areas that interest you most.

Now try these activities.

Get Active 3

a) Study the 'How to …' box on page 122.

b) Your task is to produce a *Geography in the News* sheet using the following criteria:

- You should cover at least four news articles.
- The news sheet should consist of three A4 size pages.
- Every article must be illustrated and all pictures and maps must have titles.
- There should be a list of contents on page one.
- Headlines should be given for all articles, with the largest headline for the lead article on the front page.
- All articles should be in your own words, but quotations from the source where you got the article can be included.
- Use desktop publishing software to create the news sheet.

c) When you have finished, review your work in the way that Rachel did.

Get Active 4

a) Go to Learning NI and click on the News button. In the 'Learn Newsdesk', click on 'be a reporter'. Here you will be able to read reports on international events, national or local stories that were written by young people just like you. These young people had been excited by events that were happening and just wanted others to know about them. They wanted to be reporters. Can you do the same?

b) Find an event (a piece of Geography news) that you want to write about and that you really think that others would want to read about. There are tips to help you on Learning NI. When you have written it, submit it to the Learning NI news desk. Maybe yours will be the one picked out to be published online! So get writing!

c) Now is your chance to give your views on a particular subject. Sometimes when you read a report about events you want to add your 'wee bit', agree or disagree with the view of the event as presented by the reporter. So what do you do about it? Find a story that is relevant to Geography from the Learning NI Newsdesk and email your response to it. (Use **backchat** to do this.) Have your say, and if the editor of the LNI news desk likes your work it will be published online.

Enquiry: The place where I live

Learning intentions

I am learning:

- to define what I consider to be 'my local area'
- to create a profile of my local area
- to describe my view of my local area: past and present
- how my local area might be improved in the future
- to access information from a range of resources
- to work in a group to produce a presentation.

Developing a profile of your local area

The places where we live are unique. To the people who live there they are special. People think of their 'local area' in different ways and on different scales. Some might focus on their own home. Most might include the immediate area surrounding their home, such as their road or street, section of their housing estate or village. Others might include their town or district. We are going to look more closely at what makes the places we live in special.

Aspects to consider in developing a profile of your local area
How the locality is defined: its size and grid references on the local OS map (Discoverer Series 1:50,000)
Physical or geological features: for example, is it near a river, coastline or mountains?
Areas for wildlife
Buildings and landmarks
Places where young people meet: for example, shopping centres, amusement arcades, youth centres, after-school clubs
Places for recreation: for example, leisure centres, sports clubs
Sites of historical interest or importance
What young people do in the environment: for example, litter/graffiti/vandalism, regeneration or environmental improvement schemes (wall murals)
How young people travel around
Important celebrations/events for young people
Safe and dangerous places in the local area

Get Active 1

a) Look at the table on page 125. Compile a list of aspects that are significant to the young people who live in *your* local area.

b) Describe what your local area is like today and what it was like in the past.

c) Now choose one of the following three activities.

1. What would you like your area to be like in five years' time? Draw *two* annotated sketch maps that represent an important feature (or space) in your local area, such as the main street in your town or village, your home and the area immediately around it. First, as it is today; and then as you would like it to look in five years' time:

- Think about your everyday life and what actions you could take to improve the environment in which you live.
- Ask yourself questions like: 'What do I think should happen?', 'What could I do?'
- Your sketch maps should show the location of key buildings, open or recreational spaces, roads, railways, physical features (rivers, lakes, wooded areas).
- Put titles and keys on your sketch maps.

2. Write an account of what you would like your local area to be like in five years' time.

3. Create a collage of images that show what you would like your local area to be like in five years' time.

Town scene

Park

Presenting your local area

In small groups, create a visual presentation of your local area and present this to the class. You should work with students who live in the same local area as yourself.

Tackling the presentation

a) Agree on what is meant by 'my local area'.
This is the area around your home, where:

- you meet your neighbours
- you might have attended primary school
- you may attend a place of worship
- you may play for a sports team
- you may go to a youth centre.

It is the area that you are most familiar with. Your local area is very important in shaping your identity.

b) Decide on who will be responsible for bringing together information on the different aspects of the **profile**, as outlined in the table on page 125. Each person should be clear about what they have to work on and what they have to produce.

c) Agree on *how* to present the profile. Create a resource or display that represents what is unique about the local environment as it exists for young people. Use the model shown below to do this or create your own. You should aim to include:

- visual and/or audio resources (photographs, video/audio recordings, artefacts)
- information on young people's activities and places of interest
- maps/plans or photos that show change over time (you should show maps of what your locality is like today and what you hope it will look like in the future).

Bangor, County Down, Northern Ireland

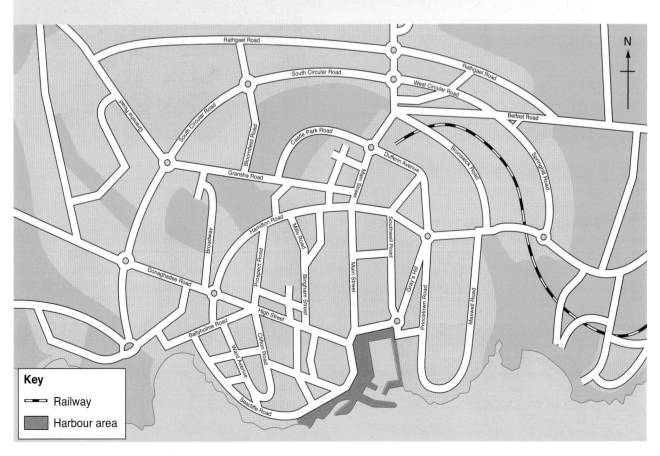

The Big Task

Creating a profile of your area

1. Draw three large concentric circles on an A3 sheet of paper or card. The three circles will be for representing your area in the past, the present and the future.

2. Label the centre as the 'PAST', the middle circle as the 'PRESENT' and the outer circle as the 'FUTURE'.

   • **Inner circle (Past)**: Put a selection of photographs (church, school, important buildings, places of work, places of recreation, and so on), maps, sketches or any other information to show what your local area was like in the **past**. Highlight the date on each resource where you can.

   • **Middle circle (Present)**: Put a selection of photographs, maps, sketches, and so on, to show what your local area is like **today**. Photocopy or draw your own map from an OS map showing your local area. Include photos of your school, church, buildings of historic interest, important shops or services, places where people work, places where young people go in their free time, nice places, quiet places, unspoiled places, dangerous places, places for wildlife, a local sports team. Include photos that say something about the local landscape, for example a photograph of a hill, wooded area, river, lake or beach. (You could take your own photographs with a digital camera.)

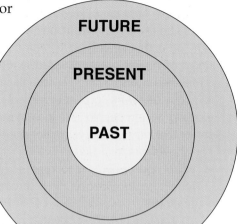

- **Outer circle (Future):** Using drawings or sketches show what some of the places and features you have included in the middle circle might look like in five to ten years' time. You might want to think about what the school of the future might look like, or a local shopping or recreational area.

3. Around the outer circle:
   - Name what you call your local area.
   - Give a 4-figure grid reference from the local OS map.
   - Give an approximate latitude and longitude.
   - Give a townland name if you know what it is and say if this name tells you anything about the area.

4. Each group will have five minutes to present its profile to the rest of the class. The class will assess each profile according to an agreed assessment criteria. The following might be assessed:
   - content of the profile
   - quality of the presentation
   - amount learned from profile and presentation.

Issues in Geography: Do we want a lignite quarry and power station in Northern Ireland?

Learning intentions

I am learning:

- how to understand a local issue
- to examine facts about an issue and make a decision
- to develop skills in working as a member of a group
- to develop research skills.

What is an issue?

An issue is something that people disagree about. A lot of Geography is about issues.

- Should a farmer turn the farm into a caravan park?
- Should an industry open up another factory in that part of a town?
- Should a new road be built across that part of countryside?

These are the kind of questions that Geographers try to answer. To understand all these issues you have to try to gather information from both sides. Don't make your mind up before you know the facts!

Facts about lignite

Before addressing this issue, we need to get some basic *facts* about **lignite** by asking some questions.

What is lignite?

Lignite can be burned like **peat** or coal. It is harder than peat, but not as hard as coal. Just like coal or peat, lignite is made of the remains of plants that grew a long time ago. It is sedimentary rock formed from layers of **decayed vegetation**.

Peat	Lignite	Coal
In Ireland, most peat is about 8,000 years old.	A piece of lignite from Ballymoney is about 2 million years old.	Most of the coal that we burn was made about 300 million years ago.
It is a very wet fuel when cut.	Lignite is wetter than coal, but not as wet as peat.	Coal is dry when the miners get it from coal mines.
It gives off very little heat. You would need to burn a lot of it to create the same heat as coal.	Lignite gives off more heat than peat, but not as much as coal.	Coal gives off a lot of heat when it is burned.

Where is lignite found?

In Northern Ireland, we know about three places where there is lignite:

- close to Crumlin in County Antrim
- close to Ardboe in County Tyrone
- close to Ballymoney in County Antrim.

They do not appear on the solid geology map on page 60 and were only recently discovered.

How much lignite is there?

In Ballymoney, geologists have worked out that there are about 700 million **tonnes** of lignite. In fact, 30 tonnes of lignite can fill a truck. So, if you were counting each truck coming out of a lignite quarry at Ballymoney at one truck each minute, it could take you over 30 years to count them all!

What can lignite be used for?

Lignite is a major **resource** for Northern Ireland because it can be used as a fuel for making electricity in a **power station**. This electricity would be used on the island of Ireland and in Scotland.

What will building the power station and quarry involve?

The land which would be quarried for lignite stretches nearly 8 kilometres from the town of Ballymoney. This is almost the whole area shown in the photograph on the right. The company wants to dig the lignite out by **open-cast mining**. It would first remove the **overburden** (the rocks and soil that lies on top of the lignite) to get at the fuel underneath. The lignite would then be burned in a power station built in the quarry.

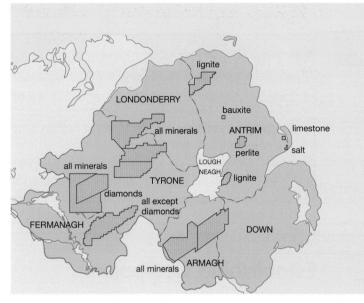

Licences to look for minerals in Northern Ireland

View of the area near Ballymoney where it is proposed to locate the quarry and power station

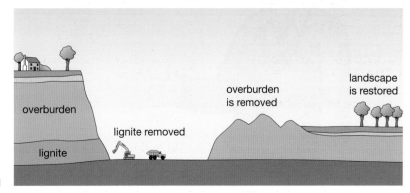

A diagram of open-cast mining

131

Get Active 1

With a partner, find out about lignite by going to www.lignite.com. Write down three things the site tells you about lignite that you did not already know.

- How can you tell if the site you are looking at is a reliable source of information? (Clue: Whose site is it? Might it be biased? Is it up to date?)
- On the map (on page 131) the lignite at Ardboe is not shown. Use an atlas to find out where it would be.
- Make a PowerPoint® presentation using images of open-cast mining and share them using a data projector. Would you want one of these close to your home? Be prepared to explain your answer. If you were looking for a job, what would be your view?

Establishing viewpoints

An issue in Geography is something that people have different views about. So, whether a lignite mine or power station should be built near to Ballymoney is a geographical issue, because:

- Those who want to dig the lignite and burn it would say:

- Others who do not want the lignite to be taken out of the ground, would say:

We need cheap electricity in Northern Ireland so that we have jobs for everybody. The quarry and power station will employ about 300 people. The soil will be put back after the lignite has been taken out – it will be even better than it was before the mining!

The lignite quarry will cover the size of 4,000 football pitches and will be 150 metres deep, deeper than 37 double-decker buses, one on top of another. Eighty farms will disappear when the quarrying destroys the land. The power station will cause pollution around Ballymoney and as far away as Scotland.

Get Active 2

a) Match the people opposite (letters) to the views they represent (numbers) from the list below:

1. Lignite is one of the most polluting fuels. We should use something else.
2. I need a well-paid job and I don't want to move to Belfast or over to England.
3. My family has lived here for hundreds of years. I want to keep it for my grandchildren.
4. We pay more for electricity than anywhere else in the United Kingdom. Businesses need cheaper electricity if they are to keep in business and keep employing people.
5. This is a rich resource which will have to be used sooner or later.
6. My house will be worth less if the quarry and power station are built.

b) Draw a concept map to show the background to, and the arguments for and against, lignite development near Ballymoney.

c) Use a search engine or an online library to find resources on lignite in Northern Ireland. Try searching 'lignite "Northern Ireland"'. Then, in small groups, use these to produce two flyers to give to people in Northern Ireland:
- To convince them that building the quarry and power station is right.
- To convince others that building the quarry and power station is not right.

d) Make up a PowerPoint® presentation to make each case (four slides for each) and, at the end, show *your* decision and give reasons for why you chose it (two slides).

e) In your group, make up a one-minute television news item on the lignite story. You will have to agree on a storyboard, a script and who is going to present it. Give the background to the lignite issue and the views of both sides. If you can, record it on video camera.

A A farmer

B A geologist

C An environmentalist

D An unemployed person

E A house owner

F An industrialist

The Big Task

1. Working with others, research another geographical issue in your area and submit it to the LNI news desk. It could be a new housing development, a new factory, a new bridge or shopping centre, or some other geographical issue. Make sure your story is fair, and try to find out what local people think about the issue. Don't make your mind up about it until you have found out all the details.
2. NIMBY is the abbreviation given to those who want developments to happen so long as they are not close to them. Find out what the abbreviation means and get some examples of people who appear to be NIMBYs.

Self-evaluation exercise

When you have completed the whole task complete the following self-evaluation exercise. Use a copy of this table. Score yourself on a range from 5 to 1 depending on how well you consider yourself to have performed.

	5	4	3	2	1	
Contributed a lot to class discussions on our issue.						Contributed very little to class discussions on our issue.
Contributed original ideas.						Had no original ideas of my own.
Showed initiative during the task.						Showed no initiative.
Showed interest and enthusiasm.						Found the whole exercise boring.
Organised my work well.						Was very disorganised.
Worked well with others.						Co-operated little with others.
Worked to the best of my ability.						Willing to do the least work possible.
Learned a lot.						Learned very little.

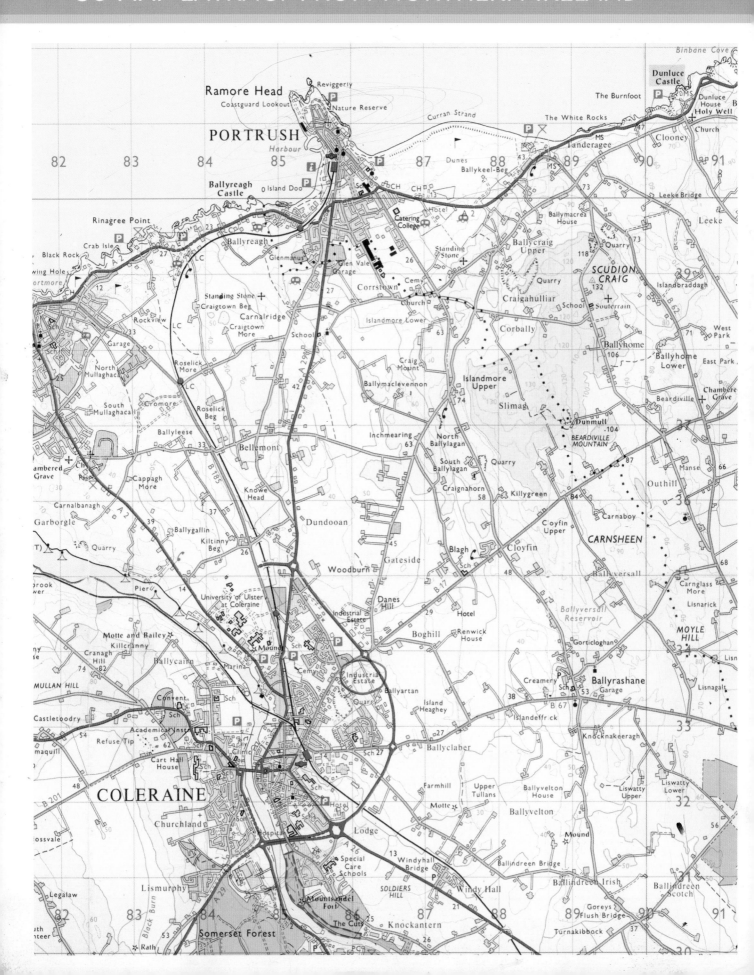

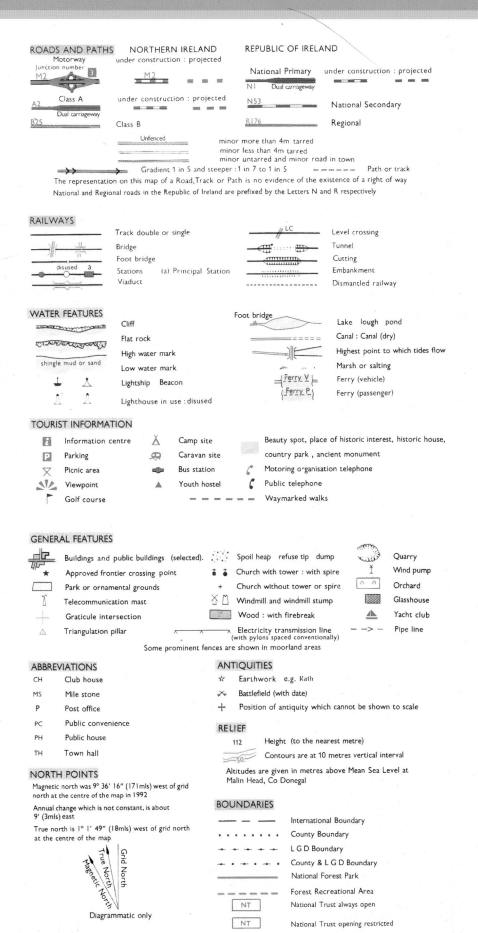

ROADS AND PATHS

NORTHERN IRELAND

Motorway
Junction number
M2   **3**

under construction : projected
M2

Class A
A2
Dual carriageway

under construction : projected

B25   Class B

Unfenced

Gradient 1 in 5 and steeper : 1 in 7 to 1 in 5   —————   Path or track

REPUBLIC OF IRELAND

National Primary   under construction : projected
N1   Dual carriageway

N53   National Secondary

R176   Regional

minor more than 4m tarred
minor less than 4m tarred
minor untarred and minor road in town

The representation on this map of a Road, Track or Path is no evidence of the existence of a right of way
National and Regional roads in the Republic of Ireland are prefixed by the Letters N and R respectively

RAILWAYS

Track double or single
Bridge
Foot bridge
disused   a   Stations   (a) Principal Station
Viaduct

LC   Level crossing
Tunnel
Cutting
Embankment
Dismantled railway

WATER FEATURES

Cliff
Flat rock
High water mark
Low water mark
shingle mud or sand
Lightship   Beacon
Lighthouse in use : disused

Foot bridge
Lake   lough   pond
Canal : Canal (dry)
Highest point to which tides flow
Marsh or salting
Ferry V   Ferry (vehicle)
Ferry P   Ferry (passenger)

TOURIST INFORMATION

Information centre
Parking
Picnic area
Viewpoint
Golf course

Camp site
Caravan site
Bus station
Youth hostel

Waymarked walks

Beauty spot, place of historic interest, historic house,
country park , ancient monument
Motoring organisation telephone
Public telephone

GENERAL FEATURES

Buildings and public buildings (selected).
Approved frontier crossing point
Park or ornamental grounds
Telecommunication mast
Graticule intersection
Triangulation pillar

Spoil heap   refuse tip   dump
Church with tower : with spire
Church without tower or spire
Windmill and windmill stump
Wood : with firebreak
Electricity transmission line
(with pylons spaced conventionally)

Quarry
Wind pump
Orchard
Glasshouse
Yacht club
– –> –   Pipe line

Some prominent fences are shown in moorland areas

ABBREVIATIONS

CH   Club house
MS   Mile stone
P   Post office
PC   Public convenience
PH   Public house
TH   Town hall

NORTH POINTS

Magnetic north was 9° 36′ 16″ (171mls) west of grid
north at the centre of the map in 1992

Annual change which is not constant, is about
9′ (3mls) east

True north is 1° 1′ 49″ (18mls) west of grid north
at the centre of the map

Grid North
True North
Magnetic North

Diagrammatic only

ANTIQUITIES

☆   Earthwork   e.g. Rath
⚔   Battlefield (with date)
+   Position of antiquity which cannot be shown to scale

RELIEF

112   Height (to the nearest metre)
Contours are at 10 metres vertical interval
Altitudes are given in metres above Mean Sea Level at
Malin Head, Co Donegal

BOUNDARIES

International Boundary
County Boundary
L G D Boundary
County & L G D Boundary
National Forest Park
Forest Recreational Area
NT   National Trust always open
NT   National Trust opening restricted

COMPILATION NOTES

Compiled from photographically reduced 1:10 000 Irish Grid material

A

adding value   using farm's land and buildings for purposes other than farming to make money, such as for tourist activities or other businesses

agriculture   using the land to grow crops and rear animals

air mass   large body of air that has blown over from somewhere else

altitude   how high up you are

arable   growing crops

assembling   putting something together, such as the parts of a car

atmosphere   layer of air surrounding the Earth

average annual   the total amount per month in a year, divided by the number of months

B

basalt   igneous rock which cooled quickly

breed   group of animals having common ancestors and certain distinguishable characteristics

C

carbon dioxide   colourless and odourless gas

carnivores   meat-eating animals

chalk   very pure form of limestone

changeable   something that changes

charter   piece of paper showing the things a group of people have agreed

chart wizard   tool to help you on the computer

chlorophyll   green substance/pigment in a plant leaf

city   large settlement which has been given city status

classify   arrange according to type or kind

climate   the average weather experienced by a particular area

climate change   any long-term significant change in the weather patterns of an area

climate graph   graph which shows both rainfall (bars) and temperature (line)

coal   black fuel made from decayed vegetation

coastal   near to or on the coast

Common Agricultural Policy (CAP)   system of agricultural support for farmers in EU countries

community   the plants and the animals that live together and make up the living part of the ecosystem

commuter   person who travels from home to work on a daily basis

concept   idea, thought or notion

concept map   diagram (sometimes called a spider diagram) where connections are made

coniferous trees   trees which bear cone-shaped fruit (such as pine, fir and larch)

consequence   something that follows naturally from an action

conservation schemes   protects land of special value for its animals or plants

consumer   people that buy goods or services; or animals that eat 'readymade food' such as plants and parts of animals

consumption   using up things

continent   one of the Earth's land masses

copyright   protection for people who have created things, so that they do not get used illegally

criteria   information about something

crop rotation   moving crops from one field to another, on a rota to keep the soil fertile

culturally diverse society   group of people who live near each other, who have many different religions, customs and traditions

D

dairy farming   farms that rear cows to produce milk

decayed vegetation   dead rotting plants

decomposers   organisms that eat dead remains of plants and animals, and then release substances that can be used by living plants

deposition   the laying down of material, for example at the bottom of the sea

developed country   see More economically developed country (MEDC)

developing country   see Less economically developed country (LEDC)

diamond rank   way of ranking things, using a frame in the shape of a diamond

diversification   using land for other activities to make money (other than the traditional farming activities)

E

earthquake   shock wave caused when plates move suddenly

economic activity   work that will earn money

economy   the money side of a community

ecosystem   community of living things and the environment in which they live

emission   something given off, such as fumes from a car exhaust

enterprise   business

entrepreneurship   the ability to start a business, usually showing considerable initiative and the ability to take some risk

environment   the area in which something exists or lives (it includes all of the living and non-living parts of that particular area)

erosion   wearing away of the Earth's surface

ethnic minority person or group of people who have a different culture, religion or language from the main one in the place or country where they live

European Union grouping of European countries (currently 27) that encourages economic links

F

fairtrade trade that pays a fair price to the producers of products

farmland land used for farming usually, in Ireland it is divided into fields by hedges

fertilisers chemicals put on the soil to help plants grow

flood warnings warnings for householders who are likely to suffer from flooding

fly tipping dumping rubbish illegally

food chain relationship between plants and animals that shows who eats what (energy is transferred from one organism to another through the food chain)

food miles measure of the distance food travels: from where it is grown, to where it is bought (or eaten) by the end-user

food web series of linked food chains showing the complex network of feeding relationships in an ecosystem

forecast guessing what will happen in the future, such as a weather forecast

fossil an imprint of a plant or animal preserved in rock

fronts where two air masses of different temperatures meet (often get clouds)

fuel something which can be burned to give off energy

G

Geologist person who studies rocks

GIS Geographical Information System (computer information)

global something that concerns the whole world

glucose type of sugar

GNP gross national product (total value of products and services produced by a country)

government people elected by voters (over 18) to run a country

granite igneous rock which cooled slowly

greywache sedimentary rock made up of larger grains in mudstone

H

habitat the area where something usually lives

hate attack any crime perceived by any person as having a racial, homophobic, religious, sectarian or disability-related motivation

hectare unit of measure equal to 2.471 acres (100 metres by 100 metres)

herbivores animals that feed only on plants

home page front page of a website

human environment the parts of an area which humans have made, such as buildings and roads

human Geography studies the behaviour of human beings, and how they interact with the natural world

I, J, K

igneous rock which has been made from molten rock which has cooled

immigrant foreigner who enters a country to settle there for a period of time

insecticides chemicals that are used to kill insects

intensive farming kind of farming that uses a lot of inputs and that may affect the environment

internet large network that links computers together all over the world

irrigated artificially supplied with water

issue something about which different people have different views

L

landscape the appearance of an area including the artificial features made by people, like roads, cities and fields

latitude distance north or south of the Equator

lava molten rock (magma) that has errupted on to the surface of the Earth

layer on top of something else

leachate liquid that has come into contact with, or has drained from waste

less economically developed country (LEDC) poorer countries (with a low gross national product GNP)

lignite fuel midway between peat and coal

limestone sedimentary rock made up of the skeletons of tiny creatures

local the area around you

M

marble metamorphic rock formed by great heat

market place where products are sold; the people who buy products

metamorphic rock which has changed from one rock to another because of great heat or pressure or both

migrant workers person who arrives in a host country (either with a job to go to or with the intention of finding one)

misperception this is misunderstood information

mixed farms   farms which grow crops and rear animals

mobile phone masts   structures with aerials for picking up mobile telephone signals

molten   rock which is so hot it has melted (also called magma)

more economically developed country (MEDC)   richer countries (with a high gross national product GNP)

mountainous   area with high land and mountains

mudstone   see Shale

N

national   something that concerns your country

nutrients   food that helps something to grow

O

omnivores   animals that eat plants and meat

open-cast mining   way of getting at rocks to take them out of the ground, if they are quite close to the surface

organic farming   type of farming that does not use artificial fertilisers and other chemical products

overburden   rock and soil which lies over the rock being quarried in open-cast mining

over production   producing too much of something

P

pastoral farms   rearing animals

peat   fuel made from decayed moss and other vegetation, common in parts of Ireland

per capita   same as 'for each head' or 'for each person'

pesticides   chemicals used to kill insects

photosynthesis   the process by which a plant produces/makes its own food

physical environment   the area around you that is of natural things (such as plants, animals, soil, rock, weather)

physical Geography   the study of the natural world: the rocks, soil, climate, rivers, plants and animals

physical map   map that shows lowlands, mountains, rivers and lakes

plate   solid part of the Earth's crust (such as the Eurasian Plate)

pollution   damage done to the natural world by humans

pollution hazards   risks associated with pollution

power station   place which burns fuel and turns the heat into electricity

precipitation   the water that falls from clouds (as rain, snow or hail)

prevailing wind   the most common direction the wind comes from

primary activity   (industry) work that is to do with getting natural resources

primary producer   someone who gets the raw materials (such as a farmer or miner)

processing   making something into something new

producer   someone who makes goods or services

producers   plants that can produce/make their own food

productive   something which produces something (in farmland it is land which can produce a lot of crops or animals)

profile   organised description of something

Q

quartzite   metamorphic rock made from a sandstone

quota   the largest amount that one is permitted to produce by regulation

R

racial incident   any incident perceived to be racist by the victim or any other person

racism   extreme dislike/hatred of a person or a group of people because of their colour, race or nationality

range   diffference between highest and lowest

recreation   how people spend their spare time

recycle   change something so that it can be useful again

recyclers   they recycle things

regeneration   transforming a rundown area with new buildings

resource   anything that can be used

retail   selling things

retailer   someone selling products to the general public

S

sandstone   sedimentary rock made up of grains of sand

satellite image   photos taken from satellites orbiting the Earth

schist   metamorphic rock caused by great heat and pressure

secondary industry   work that is to do with making things

sediment   formed by particles settling down together on the ground

sedimentary   rock which has been laid down in layers

services   job where you help or serve (like office workers, teachers or nurses)

set aside   EU scheme that pays farmers not to farm their land

sewage   what comes down our toilet pipes

shale   rock made up of mud particles (also called mudstone)

shipping agents   those who are paid to ship goods abroad

slate metamorphic rock formed after great pressure

society group of people who have something in common

sort arrange according to class or kind

source where something comes from

storyboard series of sketches and notes which are used to plan out a video or other presentation before the production begins

subsidy money given by the government to make an economic activity possible or to keep the price of a product down

surplus extra amount

sustainable something that can be continued into the future because it is not wasteful or harmful

sustainable development using resources in ways that conserves them for the future

synoptic charts maps that show weather conditions

T

tariffs amount of money to be paid in a charge or tax

technology knowledge that is used to make new products or objects

temperate mild climatic conditions (such as in Northern Ireland)

tertiary industry work that is a service to others

timber wood

tonne metric measurement (equivalent to 1,000 kilograms)

tourism people visiting an area on holiday

transportation carrying away of eroded material towards the sea

Trend the general direction in which something tends to move

Tropics an area of the world located between the Tropic of Capricorn and the Tropic of Cancer

U, V

Unconventional not usual

Vegetation plants (such as grassland, bushes or trees)

W, X, Y, Z

waste unwanted debris or rubbish

water quality amount of contamination or impurities in water

weather day-to-day changes in rainfall, temperature, sunshine

wholesaler seller of goods in large quantities (generally to a retailer who will then resell them)

workforce the people in a country who work in a job and earn money

yield the quantity produced

INDEX